Walter Kohn — Personal Stories and Anecdotes
Told by Friends and Collaborators

T0220195

**Springer**
*Berlin*
*Heidelberg*
*New York*
*Hong Kong*
*London*
*Milan*
*Paris*
*Tokyo*

**Physics and Astronomy** ONLINE LIBRARY

springeronline.com

Matthias Scheffler    Peter Weinberger (Eds.)

# Walter Kohn

Personal Stories and Anecdotes
Told by Friends and Collaborators

With 59 Figures

Springer

Matthias Scheffler

Fritz-Haber-Institut
Faradayweg 4-6
14195 Berlin, Germany
e-mail: scheffler@FHI-Berlin.mpg.de

Peter Weinberger

Technical University of Vienna
Center of Computational Material Sciences (CMS)
Getreidemarkt 9/134
1060 Wien, Austria
e-mail: pw@cms.tuwien.ac.at

Cataloging-in-Publication Data applied for
A catalog record for this book is available from the Library of Congress.

Bibliographic information published by Die Deutsche Bibliothek.
Die Deutsche Bibliothek lists this publication in the Deutsche Nationalbibliografie; detailed bibliographic data is available in the Internet at http://dnb.ddb.de.

Second printing 2004

ISBN 3-540-20809-7 Softcover edition Springer-Verlag Berlin Heidelberg New York

ISBN 3-540-00805-5 Springer-Verlag Berlin Heidelberg New York

Springer-Verlag is a part of Springer Science+Business Media

springeronline.com

© Springer-Verlag Berlin Heidelberg 2003
Printed in Germany

Typesetting: Using LATEX by editors and LE-TEX GbR, Leipzig
Cover design: E. Kirchner, Heidelberg

Printed on acid-free paper    57/3141/Ba    5 4 3 2 1 0

On the Occasion of the

80<sup>th</sup> Birthday of Walter Kohn

Walter Kohn receiving his Prize from the hands of His Majesty the King, Stockholm, December 1999 (© 1999 The Nobel Foundation, photo: Hans Mehlin)

# Preface

All of Walter Kohn's personal friends readily agree that it is almost unthinkable that he is turning 80 in March 2003. Considering all his activities – from human rights to theoretical physics – it is indeed wonderful to see that "age" has not changed his radiant lifestyle. He still seems to be one of the most frequent flyers, whether to Cuba or, as his next destination, to India and/or China, not counting his various annual transatlantic flights between Santa Barbara and Europe. He has never stopped talking to people in the same personal way, whether they are heads of state or graduate students. It is probably his open, convincing smile that brushes away all possible obstacles in communicating with fellow human beings . . .

Born in Vienna into a rather well-to-do Jewish family – his father and his uncle ran a famous art-postcards shop – he grew up in the learned atmosphere of an assimilated Jewish stratum that once carried to a large extent the cultural outburst of fin-de-siècle Vienna. There was no question that he would attend the "Akademisches Gymnasium", not because it was close by – in the same direction as the "Eislaufverein" (an ice-skating rink next to the main concert hall), where he often went on briskly cold winter afternoons – but because it was simply considered to be "the" best.

Although Austria was ruled at that time by a clerico-fascist regime whose political antisemitism created a cultural atmosphere in its image, dramatic changes in Walters's life only took place after the Nazi occupation in 1938. He was forced out of his beloved gymnasium and eventually was transferred to the (Jewish) Chajes-Gymnasium, where the number of students diminished on a daily basis. There, he also met two outstanding teachers who intellectually stimulated him for the rest of his life.

Fortunately he escaped to England on one of the last "Kindertransporte" ("children transports") – a 16-year-old boy transplanted into a foreign country who had left behind his parents. In England he was put up as an apprentice in horticulture and was later deported as "enemy alien" to Canada, where he worked as a lumberjack. But the learned atmosphere could not be banished

Walter Kohn, Thanksgiving in Santa Barbara, 2002 (photo: M. Scheffler)

even within the fenced-in camps: an ad hoc school took care of the semi-adult kids that Hitler Germany had uprooted.

Eventually he managed to be accepted at Toronto University. His first publication was written in the same spirit as his former teachers' at the Chajes Gymnasium in Vienna and dealt with contour integrations, a mathematical technique he still loves. If there was anything he rescued from the old continent, from his parents' home, something he always carried along in his mind, then surely enough it was this learned atmosphere, intellectual enlightenment and the feeling that curiosity-driven science leads to new perceptions.

This is not the place to cite or even comment on his by-now-famous papers – on the "Kohn anomalies", just one example of his various contributions to modern solid state theory, or on what is now well-known as "Density Functional Theory", for which he was awarded the Nobel prize in chemistry in 1998. Here the focus is on Walter's personal contacts – with colleagues, collaborators, friends – which were always intense, and not only with those who ended up co-authoring works with him. Some of his scientific friendships, with Luttinger or Friedel, for example, were a permanent source of new ideas. Looking back it has to be noted that – very typically for Walter – none of his friends in physics were narrow-minded scientists; all of them also had broad "other" interests.

Walter's other interests are a most rewarding aspect of meeting him, besides, of course, the notoriously sizzling discussions about certain aspects of theoretical solid state physics. It is a real pleasure to go with him to an

Mara Vishniac Kohn, summer 2002 in Berlin (photo: M. Scheffler)

exhibition, a gallery, to talk about fictional literature; his enthusiasm for the exceptionally new, for the yet-unknown old, seems to be unlimited. His enthusiasm extends from certain styles of Greek vases, discovered during a scientific trip to Crete, to drawings by Egon Schiele to be seen at a special exhibition in the capital of an Austrian province.

One cannot write about Walter without also writing about Mara, his wife, perhaps the only fixed point in the space of his vibrant activities. Mara and the house in Santa Barbara – which contains an incredible collection of precious old books and the archives of her father, Roman Vishniac, many of whose famous photographs hang on the walls – constitute the Kohn's home. The house, with its staggering view of the coastline – sometimes the Channel Islands seem to pop up only few miles away – seems an appropriate resting place for Walter before he goes off again, touring the world in matters of science or human relations. Mara and her soft, quiet voice – a friendly but decisive "Walterchen" has the ability to momentarily cut the thread of his fascinating stories about people he met or one ought to meet – occasionally give the urgently needed reminder that even a Walter Kohn needs moments of creative quietness.

Quite a few stories collected in this book give account of the circumstances of working with Walter, of being a collaborator of his, of being accepted as a friend of the Kohns. There is no doubt that these stories also cover a certain chapter in the history of science in a very personal manner: a kind of "oral history" of modern theoretical physics. Besides physics, besides stories about physics, however, they all speak to Walter Kohn's enduring humanity – if only by recalling his infectious, giggling laugh, which he must have saved from his early childhood.

Berlin, Vienna,                                                                    *Matthias Scheffler*
January 2003                                                                      *Peter Weinberger*

**About the Editors**

**Matthias Scheffler** is director of the Theory Department of the Fritz-Haber-Institut der Max-Planck-Gesellschaft and professor at the Technical University and at the Free University Berlin. His research concerns condensed-matter theory, materials, and the chemical physics of surfaces. His current interests include developing first-principles methods (using density-functional theory) for molecular simulations that bridge the time and length scales from those of the atomistic processes to those that determine the properties of realistic systems. Contact: Fritz-Haber-Institut, Faradayweg 4-6, D-14195 Berlin-Dahlem, Germany; scheffler@fhi-berlin.mpg.de; www.fhi-berlin.mpg.de/th/th.html

**Peter Weinberger** is acting director of the Center for Computational Materials Science (CMS) and professor at the Technical University of Vienna. His main scientific interests are directed to condensed matter theory, in particular to the theory of transport (electric, magneto-optical) in systems with "reduced" translational symmetry (layered systems, nanostructures, etc.) and to time-dependent spectroscopies. He is also author of several "non-scientific" books (fiction: novels, short stories). Contact: CMS, Getreidemarkt 9/134, A-1060 Wien, Austria; pw@cms.tuwien.ac.at; www.cms.tuwien.ac.at

# Contents

**Walter Kohn as Ph.D. Advisor**
*Vinay Ambegaokar* ............................................... 1

**Memories of Walter Kohn at Bell Labs and After**
*Philip W. Anderson* .............................................. 3

**Walter Kohn: My Ideal of a "Man of Science"**
*Wanda Andreoni* ................................................. 6

**A Twinkle in the Eyes**
*Neil W. Ashcroft* ................................................ 10

**The Max von Laue Kolloquium
at the *Physikalische Gesellschaft zu Berlin***
*Klaus Baberschke* ............................................... 12

**More Than a Founding Father**
*Giovanni Bachelet* .............................................. 15

**Walter Kohn, the Chemist**
*Evert Jan Baerends* ............................................. 20

**Looking 50 Years Back:
Walter, on the Occasion of His 80th Birthday**
*Sidney Borowitz* ................................................ 23

**Meeting Walter Kohn**
*Ewa and Krzysztof Brocławik* .................................... 26

**Of Sandboxes and Silences**
*Kieron Burke* ................................................... 30

**A Glimpse of Walter Kohn**
*Sudip Chakravarty* .............................................. 32

Contents

**Quantum Reflections**
Dennis P. Clougherty............................................. 35

**Not a Relative, but ...**
Marvin L. Cohen ................................................ 39

**Riding Walter's Wake**
Morrel H. Cohen................................................. 42

**Paris Acquaintances**
Claude Cohen-Tannoudji.......................................... 47

**Walter Kohn and the UCSB Campus**
France Anne Córdova............................................. 48

**Principle and Practice**
Bernard Delley .................................................. 50

**Thanksgiving at the Kohns**
Jacques des Cloizeaux ........................................... 52

**Walter Kohn and UC Policy**
Hugh DeWitt..................................................... 54

**Walter Kohn and the Australian Psyche**
John Dobson .................................................... 57

**Reunion in History**
Gertrude Ehrlich................................................. 60

**For Rappa on His 80ᵗʰ Birthday from Terry**
Josef Eisinger .................................................. 63

**Meeting Walter Kohn in Internment Camp**
Ernest L. Eliel.................................................. 66

**Meeting Walter Kohn**
Gerhard Ertl ................................................... 68

**Impressions of Walter Kohn**
Helmut Eschrig ................................................. 69

**Reminiscences of Walter**
Peter J. Feibelman.............................................. 72

**To Know How to Count ...**
Michael E. Flatté .............................................. 74

**My Friend and Colleague Walter Kohn**
Jacques Friedel ................................................ 76

**Working with Walter**
*Michael Geller* ............................................... 78

**A Chinese Portrait of Walter Kohn**
*Antoine Georges* ............................................. 81

**Memories of Walter Kohn in Santa Barbara**
*Hardy Groß* ................................................. 82

**Thinking Back**
*Werner Hanke* ............................................... 85

**Thank You Walter for All You Have Taught Us**
*Michael J. Harrison* ......................................... 89

**Vision Realized in Details: To Walter on His 80<sup>th</sup> Birthday**
*James B. Hartle* ............................................. 91

**Walter Kohn in Japan**
*Hiroshi Hasegawa* ........................................... 93

**My Favorite "Walter Story"**
*Alan J. Heeger* .............................................. 96

**Walter Kohn as a Scientist and a Citizen**
*Conyers Herring* ............................................. 97

**A Personal Tribute to Walter Kohn on His 80<sup>th</sup> Birthday**
*Pierre C. Hohenberg* ......................................... 99

**Recollections of a Long Friendship**
*Daniel Hone* ............................................... 103

**My Recollections of Professor Walter Kohn**
*Naoki Iwamoto* ............................................. 106

**With Walter: 40 Years of Friendship**
*Denis Jérome* ............................................... 109

**The Noble Day**
*Alex Kamenev* .............................................. 112

**The Pleasure of Getting to Know Walter Kohn**
*Boris Kayser* ............................................... 115

**Happy Birthday Walter!**
*Roland Ketzmerick* .......................................... 116

**Memories of a Great Scientist and a Great Person**
*Barry M. Klein* ............................................. 118

Contents

## Many Happy Returns, Walter!
Norman Kroll ...................................................... 120

## A Postdoc with Walter
Norton D. Lang ..................................................... 122

## Reminiscences on the Occasion of Walter Kohn's 80$^{th}$ Birthday
James S. Langer .................................................... 124

## Some Recollections About Walter and His Work
David Langreth ..................................................... 127

## Romancing the Theorem
Mel Levy ........................................................... 132

## Crete, June 1998
Michele Levy ....................................................... 135

## Recollections of Walter Kohn
Steven G. Louie .................................................... 138

## Walter Kohn – Points of Impact
Bengt Lundqvist .................................................... 140

## KKR – Reminiscences About Walter Kohn
Gerald Mahan ....................................................... 144

## Cecam's Visitor!
Michel Mareschal ................................................... 146

## Nobel Mania
Ann E. Mattsson .................................................... 148

## Walter's Contagious Conscience
Eric McFarland ..................................................... 153

## Memorable Moments with Walter Kohn
N. David Mermin .................................................... 155

## Impressions of Walter Kohn
Horia Metiu ........................................................ 160

## ... Small Matters
Douglas L. Mills ................................................... 165

## Walter Kohn in Cracow
Roman F. Nalewajski ................................................ 167

**Walter Kohn**
*Venkatesh Narayanamurti* ........................................... 171

**A Class with Class**
*Herbert Neuhaus* ................................................... 173

**Great Influences in a Small Country**
*Risto Nieminen* .................................................... 177

**Walter's Group Parties**
*Qian Niu* .......................................................... 181

**He Did Effective-Mass Theory Too!**
*Sokrates T. Pantelides* ............................................. 183

**Thoughts and Recollections
of a Remarkable Man**
*Dimitrios A. Papaconstantopoulos* ................................. 185

**The Bonding of Quantum Physics with Quantum Chemistry**
*Robert G. Parr* .................................................... 187

**My Meetings with Walter Kohn**
*Michele Parrinello* ................................................. 192

**The Greatest Gift Is Something Worth Thinking About**
*John P. Perdew* .................................................... 194

**Vignettes: Switzerland, Australia, Santa Barbara**
*Warren E. Pickett* ................................................. 196

**Theory Versus Reality**
*John J. Rehr* ...................................................... 198

**Recollections of Walter in La Jolla and Zurich**
*Maurice Rice* ...................................................... 200

**I am Happy that the R Stands for Rostoker**
*Norman Rostoker* .................................................. 206

**It Started with Image Charges**
*Joseph Rudnick* .................................................... 208

**A Math Teacher's Little Poem**
*George Sanger* ..................................................... 211

**Stories About Walter**
*Andreas Savin* ..................................................... 212

Contents

**A Special Reunion**
*Douglas Scalapino* . . . . . . . . . . . . . . . . . . . . . . . . . . . . . . . . . . . 214

**Walking and Talking in Berlin**
*Matthias Scheffler* . . . . . . . . . . . . . . . . . . . . . . . . . . . . . . . . . . . 216

**My Personal Walter Kohn Story**
*Sheldon Schultz* . . . . . . . . . . . . . . . . . . . . . . . . . . . . . . . . . . . . 224

**Walter Kohn and Vienna**
*Karlheinz Schwarz* . . . . . . . . . . . . . . . . . . . . . . . . . . . . . . . . . . . 227

**We Met at the Institute in Copenhagen**
*Benjamin Segall* . . . . . . . . . . . . . . . . . . . . . . . . . . . . . . . . . . . . 231

**Happy Birthday, Walter**
*Lu J. Sham* . . . . . . . . . . . . . . . . . . . . . . . . . . . . . . . . . . . . . . . 234

**A Mean Martini**
*David Sherrington* . . . . . . . . . . . . . . . . . . . . . . . . . . . . . . . . . . . 238

**Lunch with Walter**
*Mark E. Sherwin* . . . . . . . . . . . . . . . . . . . . . . . . . . . . . . . . . . . . 241

**Act of Compassion**
*Bonnie Scott Sivers* . . . . . . . . . . . . . . . . . . . . . . . . . . . . . . . . . . 244

**Some Recollections of Life with Walter Kohn**
*John R. Smith* . . . . . . . . . . . . . . . . . . . . . . . . . . . . . . . . . . . . . 246

**Bonjour Mon Très Cher Ami
et Bonne Anniversaire**
*Charles Sommers* . . . . . . . . . . . . . . . . . . . . . . . . . . . . . . . . . . . . 249

**Walter Kohn and Boris Regal: The Early Days of ITP**
*Robert Sugar* . . . . . . . . . . . . . . . . . . . . . . . . . . . . . . . . . . . . . . 252

**Flashback to My Post-doc Days with Walter**
*Yasutami Takada* . . . . . . . . . . . . . . . . . . . . . . . . . . . . . . . . . . . 257

**A Challenge to My Own Scientific Thoughts**
*Walter Thirring* . . . . . . . . . . . . . . . . . . . . . . . . . . . . . . . . . . . . 260

**Walter Kohn's Influence on One Engineer**
*Matthew Tirrell* . . . . . . . . . . . . . . . . . . . . . . . . . . . . . . . . . . . . 262

**A Tribute to Walter Kohn**
*Ulf von Barth* . . . . . . . . . . . . . . . . . . . . . . . . . . . . . . . . . . . . . 264

Encounters with Walter:
From the Mysterious Second "K" to the Local Secretary
*Peter Weinberger* .............................................. 267

An Encounter with Walter Kohn in Orsay
*Jerry L. Whitten* .............................................. 273

Walter Kohn, World Citizen and Professor Extraordinaire
*Henry T. Yang* ............................................... 275

Meeting with Walter Kohn
*Weitao Yang* ................................................. 280

My Experience of Working with Walter
*See-Chen Ying* ............................................... 282

Walter Kohn: Mentor and Role Model
*Eugene Zaremba* ............................................. 285

The Scientist and Human Values
*Joseph Zycinski* .............................................. 288

Appendix I – Autobiography
*Walter Kohn* ................................................. 293

Appendix II – A Musical CV
*David Mermin* ............................................... 307

# Walter Kohn as Ph.D. Advisor

Vinay Ambegaokar

Cornell University, Ithaca, U.S.A.

Much of life is luck. A member of the class of 1955 in Mechanical Engineering at MIT, I discovered that, because I had talked my way out of English and Chemistry, and worked hard, they would give me a Master's degree for staying an extra semester. During those months, in addition to doing some forgettable research on the creep of metals, I took half a course on quantum mechanics – taught, as it turned out, by Sidney Drell – thinking that a slightly better understanding of Nature could not hurt a practicing engineer. It did more. Before leaving Cambridge, to return to India and find work, I applied to some graduate schools for admission in Physics and asked Drell for advice. He told me about a Harvard instructor named Walter Kohn who had recently gone to Carnegie Tech, and was doing interesting work in solid state physics. During the summer of 1956, as I contemplated jobs in which the salary was inversely proportional to the challenge, telegrams arrived offering me fellowships to Harvard and Carnegie. I accepted the latter, causing consternation among my American friends' mothers.

So it came about that at 22, I met an oh so mature, pipe-smoking, wise, research scientist – aged, I now calculate, 33! (Only later did I discover another side to him, a side that could be reduced to uncontrollable tear-making merriment over something like a poem by Christian Morgenstern.) Gravely, he advised me about courses, including the wonderful suggestion that I might profit from a second course in quantum mechanics taught by Gian Carlo Wick, even though I had not finished a first one. The second year was a disappointment; he was on leave and the courses on statistical mechanics and solid state physics, which I should have taken from him, were taught badly and bizarrely by someone else. By contrast, the last year and a half were quite wonderful. His suggestion for a research project was very much to my taste, and his supervision of it was precise and constructive. He thought hard during our meetings about how to keep the project moving along. A steady stream of seminar visitors, including the likes of John Bardeen talking impenetrably about the new theory of superconductivity, made it clear that, largely because of Walter, I was at a major research center for my new line of work. He arranged for me

1

to spend the summer of 1959 at the Bell Labs, then the ultimate ivory tower and a summer paradise for solid state physics. ["...les vrais paradis sont les paradis qu'on a perdus." – Proust.] There was a semester visit by John Ward, whose course gave me a liberal education in how to calculate deftly .... And, all too soon, it was over, though not before Walter helped me find a post doctoral berth in Copenhagen, and then a position at Cornell.

My wife Saga and I remember Lois and Walter Kohn very fondly from those years. There were many social occasions, thanksgiving dinners, tennis parties, and opportunities to become easy with each other.

Owing so much to him, I offer these old memories in friendship on his 80$^{th}$ birthday.

**About the Author**: Vinay Ambegaokar is Goldwin Smith Professor of Physics at Cornell University. His Ph.D. thesis was on a many body formulation of the effective mass theory of semi-conductors with few carriers, building on Walter Kohn's pioneering work. Since then he has worked on various problems in condensed matter and low temperature physics, and written a book "Reasoning about Luck: Probability and its uses in physics" (Cambridge University Press, 1996) for serious general readers. Contact: Department of Physics, Cornell University, Ithaca, NY 14853, U.S.A.; va14@cornell.edu

# Memories of Walter Kohn
# at Bell Labs and After

Philip W. Anderson

Princeton University, Princeton, U.S.A.

Walter was one of an exciting generation of students who showed up at Harvard right after the war, having all kinds of interesting backgrounds – Los Alamos and other wartime labs, the military, refugees like Walter, mixed in with very young recent undergraduates For good reason, he took his studies very seriously, and he may have thought of me as rather light-weight – in any case, we were not much more than acquainted, and went our separate ways, he to a postdoc in Europe and I to Bell Labs, having now become serious, since I had acquired a family to support. So my very real and deep friendship with Walter began after '53, when he and Quin Luttinger began the annual summer visits to Bell Labs which were so enormously fruitful for both parties. They represented for us an important contact with the wider world of theoretical physics as it was then evolving, with new mathematical methods such as quantum field theory, as well as a new lifestyle which Jeremy Bernstein was to later dub "the Leisure of the Theory Classes" involving personally administered research contracts which allowed lecturing at summer schools or workshops held in glamorous resorts, attendance at international meetings, and the like. But at the same time Quin and Walter were producing a series of papers of great practical as well as intellectual value, doing such things as sorting out effective mass theory in the complex case of real semiconductors, putting transport theory on a sound basis, and comparing detailed theory of shallow impurity states with experiment. Walter, in particular, wrote several of the papers I consider his best during that period, papers obviously stimulated by the Bell group and Bell's experimentalists' needs.

It is Walter's great strength to find important new physics in areas which ordinary people consider to have been long since mined out. Such a paper was his discussion of the localizability properties of Wannier functions, which he worked out in one dimension but in such a way that Blount could base a comprehensive review on his methods. His remarkable paper revisiting the question of why insulators are so to all orders, and introducing the $U(1)$ gauge symmetry, is a prime example. And his paper on the "$n + 1$ body problem" was an important precursor of Landau theory.

3

Our appreciation of Walter was not diminished by the fact that we were able to use him as a stick with which to beat the Bell Labs management, which, if possible, appreciated him even more than we did. In 1955–56 we (Conyers Herring, Peter Wolff, as well as myself, and at the beginning Harold Lewis and Gregory Wannier) were encouraged to organize a new theory department, probably because of the rather rapid attrition of theorists from the Labs. We had of course lost Charlie Kittel and John Bardeen, and both Harold and Gregory were negotiating transfers to academia at the time. One of the commonest phrases we used in the negotiations with our managers was "let's organize a department into which Walter Kohn would be happy to come". We managed to get, with this as our strongest argument, a democratic structure with rotating chairmanship, postdocs, of course a measure of control over our own fates (until then we had been separately supervised by experimental department heads), a liberal travel and sabbatical policy, and (quietly and informally) considerably improved salaries – a lot to owe Walter for!

I am reminded also of another almost forgotten cause for our appreciation. Bell Labs had its own incident of McCarthyism, a "questionnaire" about one's links, past and present, with "subversive" organisations, which was required of all employees. Why Walter as a temporary was asked to respond, I don't know, but I do know that he refused, providing a measure of cover for the shockingly few permanent employees who refused to fill it out as a matter of principle (Gregory, Alan Holden, perhaps Hal Lewis, and myself, among my immediate acquaintance). There were no consequences – this was late in the McCarthy era, true to the AT&T's characteristic ponderousness, and by the time the responses were all in, my reaction would have looked silly. But we did not know that at the time. Walter and Alan Holden, with their mutual interests in music, the arts, and literature, became good friends, an additional link from Walter to Joyce and myself.

Of course, we should be so lucky – we did not attract Walter, instead he decided to take on the responsibility for growing the physics department at the new branch of the University of California at San Diego – an assignment he carried on with remarkable success, as I'm sure is detailed elsewhere in this book. But perhaps others will not realize one of the major sources of that success, namely his intimate knowledge of Bell Labs personnel, gained over many years of summer visits. Three of the mainstays of the fledgling department were Bernd Matthias, Harry Suhl and George Feher, three of Bell's most exciting and original finds. So, over the years, I feel that Walter and Bell came out about even – although that's not counting his postdocs and students, like Maurice Rice and Pierre Hohenberg, who came our way.

Of course, Joyce and I remained friends with Walter and his then wife; we visited San Diego a number of times, notably in 1967 when Bernd and he brought me there for a month as Regents' lecturer, and I gave the talk that became "More is Different". But we seemed to drift into different branches of physics until he became head of the ITP, where I labored for three years as a member of his advisory board during his tenure; and since then I have taken great pleasure, on visits to the ITP, in seeing him and Mara in the wonderful setting of their hillside house filled with memorabilia from her family – I always try to sneak into that wonderful library and snoop about during any visit.

**About the Author**: Philip Anderson is Joseph Henry professor, emeritus, in the physics department of Princeton University. His interests at present are (1) the theory of strongly correlated electron systems, particularly high-$T_c$ superconductors; (2) aspects of complexity theory, as a long-time steering committee member of the Santa Fe Institute, and one of the founders of its economics program; (3) reviewing popular books on science, mostly for the Times (London) Higher Ed Supplement, and occasionally writing on such subjects. He won 1/3 of the Nobel Prize in physics in 1977; his most recent honorary degree is from the University of Tokyo, Dec 2002. Contact: Dept. of Physics, Princeton University, Princeton, NJ 08544, U.S.A.; pwa@pupgg.princeton.edu; pupgg.princeton.edu/www/jh/pwa/

# Walter Kohn: My Ideal of a "Man of Science"

Wanda Andreoni

IBM Research Division, Zurich Research Laboratory, Rüschlikon, Switzerland

It was a hot summer in Paris in 1974. My small dusty room in the pension at rue Madame was suffocating at night time and the two spinsters were distributing only rancid bread for breakfast but I was living a special moment of my life which I would always remember. Walter Kohn was visiting CECAM for a month and I had the privilege to work under his guidance. I will never forget that morning when Carl Moser entered my office followed by his grunting dogs, sat awkwardly on the chair of my officemate and announced with that strident voice of his "Wandá ... Walter is coming". Finally it made sense for me to have left Rome, my family and my colleagues (who were also my friends) at the University, to come and land in that funny place devoid of books and scientific magazines, where people were coming and going often not even talking to each other, where the only useful collaborator was an IBM computer. I knew Walter Kohn only from solid state physics textbooks and from some of his articles which I kept aside as precious things. I was working on a code that would calculate Wannier functions using his recently published method. Magically one summer day that famous name had materialized: Walter Kohn had come indeed. I had not wasted time figuring out how he would look like but still my encounter with him was a surprise. Walter was a kind person, as too few scientists are. In contrast with many of his much less famous colleagues, he did not make you feel like a nobody. For me it was also a surprise that contrary to many theoreticians of those times – the times of the decks of punched cards – Walter Kohn was genuinely interested in computational sciences and valued the uniqueness of the computer as a research tool.

The time I was spending with him was precious to me. Walter was not only important for my education as a physicist but for my development as a person. He cared to give me advice on my personal life, with the discretion and elegance that have always characterized his approach to any matter. I will always remember the day when I declined his invitation to have dinner with A. Blandin adducing as a pretext the fact that I wanted to have some new results for the day after – which was also true but not the main reason for my obstinate refusal. I felt too little to be with him. Respect is too a weak word to describe what I felt for Walter. It was reverence. I remember that nice summer evening when he insisted on introducing me to Asian food. We went to a Vietnamese restaurant in the Quartier Latin together with Priya Vashishta; Priya would remind me of it many years later in San Francisco. I remember being with him sometimes when he was taking a walk after lunch in Orsay, a habit that he had recently started and kept since then. It was during one of those little walks in Orsay that Walter told me about the time when he moved to La Jolla, and of how the view of Sorrento reminded him of the first sight he'd had of the cliffs of South California. Whatever Walter was talking about, he had something non-trivial to say, original, interesting. I always felt I had learnt something from staying with him. Rarely in my life would I feel the same in the company of a scientist.

I lost contact with Walter during the subsequent years occasionally. Many are indeed the memories that are coming back to my mind: the occasions when I met Walter are intermixed with the development of my life and my moves from one place to the other, France, Italy, US and Switzerland. I was still in Lausanne when, in 1985, the opportunity of organizing an international scientific event was given to me. It was a summer school on a Swiss mountain resort, for which I had found a title that at the time was not so common: "Electron states in atoms, molecules and solids: the point of view of the physicist and of the chemist." What better than density-functional theory as a link between physics and chemistry? And who better than Walter could ever have taught density-functional theory? I decided to invite him. To the amazement of most of the local people who had reacted to my suggestion with a ... "Wandá ... tu as dit Walter Kohn ? le grand bonhomme ? mais ... il ne va pas venir, tu es folle ... ", Walter gladly accepted the invitation. It was a real joy to have Walter at Les Diablerets teaching density-functional theory to a heterogeneous and enthusiastic group of people: graduate students, junior and senior researchers, both physicists and chemists. He was happy, relaxed, enjoying the sunny air of the Swiss mountains and available for anybody who was eager to learn.

From that time I have had a regular contact with Walter. My move to IBM Zurich a year later helped also because Walter comes every other year to visit Maurice Rice at the ETH. Walter has continued to teach me more than physics. He has saved me from my frustration at being unable to learn German the way I wanted, namely by reading writers like Franz Kafka and Hermann Hesse. I did not have the patience for either following a boring course or reading a book with a dictionary at hand. Walter suggested me to buy those books having the original text on the left and an Italian or French or English translation on the right. Provided I was not tempted to just read the right side of the book, I would learn. And I did. It was Walter who told me how to take advantage of the good program of concerts at the Tonhalle without having to make a reservation in advance, by simply going there when

I felt like it with a light heart, no expectations and seeing whether any seats were available.

It was Walter who during the difficult times I have had in my personal life, has found words of encouragement that have made a difference for me. Maybe others may have told me similar words but obviously when they came from him I did not forget them. My memory goes back to the day (October 18, '95) when I had an appointment with Walter at ETH and he asked me what was going on with me. I must have looked particularly sad. It was the day when some tribunal in England was deciding on my divorce irreversibly. Walter had been following the absurd stories in Zurich which had originally forced my husband to leave home, and having such a friend near me that day was a gift. My memory goes back to the day (in January 2001) when I came back to my e-mail after days spent in the hospital where my mother had been brought after having a stroke. I felt better when I saw in my reader list a Happy New Year message from Walter, and he certainly did not neglect to send me his support later on.

A very nice recent memory is that day in Catania. It was June '99. From an uncertain rainy spring in Zurich I had been pleasantly transported into a warm summer, full of the perfumed breeze and the vivid colors of the Sicilian coast. On the occasion of the annual meeting of the National Institute of the Physics of Matter, Carlo Calandra, president and organizer had invited a few selected speakers from abroad. Walter and I were on this list and we were also in the same Hotel. I stayed there only one day and in the evening we had dinner in one of those typical small restaurants facing the sea. It was there that Walter told me about his young years in Vienna and about his love and talent for Latin. It was a pleasant discovery for me, which confirmed my conviction that somebody with a scientific mind cannot help being attracted by the beauty of the Latin language and that a humanistic culture is fundamental to the ability of doing great science.

I have known Walter for 29 years. What I feel for him, whenever he calls me from ETH to announce his presence in Zurich, has not changed since I first met him in Orsay. I feel shy, my English starts to become confused but I am happy, happy that he has not forgotten me and that I will see him soon. While most people I considered with admiration when I was young have disappointed my expectations because they revealed to be narrow minded and humanly mediocre, the sentiment I have for Walter is untouched.

Walter remains a towering example for me of what a great scientist should be: a man of science. The term "scientist", coined around 1840 to simply designate "the practitioner of science", is too little for Walter. When we are adolescents we imagine "scientists" as people who are superior, bright both in their minds and in their hearts, because we tend to assimilate deep knowledge and insight with high standards in ethics and human values. In our dreams, we are tied to the renaissance ideal of a man of culture and also to the belief that "pietas" is a characteristic of any man of culture. It takes some time to realize that reality is dramatically different, that successful "scientists" may be as mean, selfish and narrow as anybody else, that their drive is often mere personal ambition rather than genuine curiosity and fascination for nature and that behind some intelligent approach to the laws of nature there is often full ignorance of art, music, literature and history, sometimes even contempt for all those disciplines that are not scientific.

Fortunately, however, there are a few who stand out of the grey reality of our world and shine like the stars of our dreams: this is what Walter Kohn is for me.

**About the Author**: Wanda Andreoni manages Computational Biochemistry and Materials Science at the IBM Research Laboratory in Zurich. Before joining IBM Research in 1986, she has been working in several other places: at CECAM-Orsay where she first met Walter Kohn, at the University of Rome (Italy), at the Ecole Polytechnique Federale in Lausanne (Switzerland), at the University of Geneva and has visited for extended periods Bell Laboratories and Exxon Research. She has been working on diverse applications of density-functional theory, from semiconductor physics to cluster chemistry to biochemistry and food chemistry. Contact: IBM Research Division, Zurich Research Laboratory, CH-8803 Rüschlikon, Switzerland; and@zurich.ibm.com

# A Twinkle in the Eyes

Neil W. Ashcroft

Cornell University, Ithaca, U.S.A.

My memory is becoming increasingly porous, but I believe I first met Walter in February of 1965, in San Diego, during a short reprieve from the winter besetting the University of Chicago (and I deduce from this that Walter must have been on the right side of 40 then). At Cambridge, where I had studied for my Ph.D., I had made very close acquaintance with the metallic state of matter. But the insulating state, in a fully many-body context, was a puzzle to me then. Walter had written some beautiful papers on this topic, and in early 1965 began to set me on the right path. I thought then (and do so now) that he was one of the most congenial of men, and that he was endowed, as we all know, with the most extraordinary physical insight and perspicacity. The nearly four decades that have since passed have only served to solidify my view; in fact during this period I simply cannot count the number of times that, tracking back through the literature on a particular research topic in condensed matter theory, I have ended up at a seminal paper of Walter's. It is abundantly clear that he has had a profound impact not only on our chosen field, but on allied fields as well. His remarkable creation of the density functional approach to correlated quantum systems has actually far transcended theoretical condensed matter physics per se.

I was made aware of this, in a rather amusing way, in October of 1998, on one of those especially luminous days when the Nobel Prizes are announced. By a strange conjunction of world lines, I happened to be attending, at the University of California, Santa Barbara, a meeting of Directors of NSF's Materials Research Centers. There was much rejoicing, as I recall, at the morning's news on that very day that Walter was co-winner of the Nobel Prize in Chemistry. At a mid-afternoon break, I thought it at least worth a try to see Walter and to offer my congratulations personally, though given the boisterous activity at Broida Hall I assessed my chances as slim. But by a remarkable fluctuation there he was, in his office, momentarily unencumbered.

Walter was in high spirits as might well be expected and we had an entirely vivifying chat, mainly about the long story of the development of electronic structure. But at last I had to get back to render unto Caesar, and just as I was leaving Walter said, with a slight frown framing nevertheless a clear twinkle in the eyes, "But Neil, what will the physicists really think of all this?" with a distinct emphasis on 'physicists.' Well, I did not think it my place to speak for the grand-ensemble of all physicists, but having a somewhat ecumenical, and perhaps even a 'colonial' viewpoint on these matters I answered Walter that I thought there was actually a pretty good precedent. One Ernest Rutherford had also received the Nobel Prize in Chemistry.

As we all know, to our common delight and long-term pleasure, Walter has a truly encyclopedic memory and there is also not much in the history of physics that has escaped his attention. He quickly responded with the most potently well-justified and yet 'politically correct' of all possible answers: "Yes, and Marie Curie too".

Walter's work has meant a great deal to me, actually more so on average in the classical domain than in the quantal. For classical systems where strong correlation is a sine-qua-non (for example in the burgeoning field of inhomogeneous classical fluids) the density functional viewpoint has led to quite remarkable advances, many of them stemming from appropriate classical limits of the quantum problem and stemming from Walter's insights. He is surely one of the great physicists of our age, and at this 80 year mark I thank him for his many, many contributions that have so enlivened our subject.

**About the Author**: Neil W. Ashcroft received his Ph.D. from Cambridge University in 1964, spent a year as a post-doctoral associate at the University of Chicago, and has been at Cornell University (where he is now Horace White Professor of Physics) ever since. Contact: Physics Department, Clark Hall, Cornell University, Ithaca, NY 14853-2501, U.S.A.; nwa@ccmr.cornell.edu

# The Max von Laue Kolloquium
# at the *Physikalische Gesellschaft zu Berlin*

Klaus Baberschke

Freie Universität Berlin, Berlin, Germany

The Physikalische Gesellschaft has a tradition of organizing twice a year the prestigious Max von Laue Kolloquium. The PGzB is more than 150 years old – it was founded in 1845. In 1951 and 1955 Max von Laue was its president. Later the society created this series of lectures.

Reception in the library of the physics department after the MvL Kolloquium on 9th November 2001 (photo: Physikalische Gesellschaft zu Berlin)

Being in charge of the PGzB in 2000/1 it was a great pleasure for me and Karl Bennemann to convince Walter Kohn to give this lecture here in November 2001. His talk about " ... Wave Functions and Density Functionals ... " was a great success. Several hundreds came – he simply doubled, if not tripled, the number of people in the audience, as can bee seen in Fig. 1 at the reception in our library. Colleagues and many students from all over, the

Humboldt, the Technical, and the Free University as well as from Potsdam came. Walter was pleased in particular that he could welcome in the audience a former co-worker of his who just became a new member of our department – Hardy Groß. He also met many old friends from the early days in La Jolla, like Ursula and Klaus-Dieter Schotte. Almost equally important as the enthusiastic lecture was the lively discussion in the library, which went on until late at night. During such a long discussion Walter obviously needed to fortify himself (see Fig. 2): *"Essen und Trinken hält Leib und Seele zusammen"*.

Some refreshments in the library (photo: Physikalische Gesellschaft zu Berlin)

For such an honored guest it was not difficult to get a special private guided tour at the *Jüdisches Museum* (Daniel Libeskind, architect) the next day. We spent the full morning there and all of us were deeply impressed. During the tour through the museum we speculated that Walter would like to come back, bringing with him his wife Mara to show her the *"Neue Berlin"*. That meanwhile has taken place, when they visited Matthias Scheffler at the Fritz-Haber Institute in spring 2002. For the future, we wish both of them all the best and would like to have them back here in Berlin many more times.

**About the Author**: Klaus Baberschke is professor of physics at the physics department, Freie Universität, Berlin. His research concerns solid state and surface physics, in particular magnetism. In his current interests he uses ferromagnets of few atomic layers only, to manipulate the Curie temperature, study spin fluctuations in *2D*, critical phenomena, and induced magnetism at interfaces. Contact: Freie Universität Berlin, Institut für Experimentalphysik, Arnimallee 14, D-14195

Klaus Baberschke

Berlin-Dahlem, Germany; bab@physik.fu-berlin.de; www.physik.fu-berlin.de/~ag-baberschke/index.html

# More Than a Founding Father

Giovanni Bachelet

Università La Sapienza, Rome, Italy

## Walter Kohn in Flesh and Bone

I first met Walter Kohn during the summer of 1994, when I attended the ITP program on Quantum Many-Body Computations for Condensed Matter Physics, at the invitation of David Ceperley, one of its Directors. It is always a great emotion to meet a Founding Father of your field of research at long last. More so if the meeting venue happens to be named after him – the Kohn Hall. You may not realize that, but you secretly expect a wise man with a long white beard and a halo shining above his head; or at least, extrapolating from previous experiences with much less prominent colleagues, an infinitely busy and self-conscious person who will barely notice your presence. Well, Walter's look and his way of attending seminars and meeting people were antipodal with respect to these expectations: he would appear genuinely interested in people and arguments, and would often be the first to nicely introduce himself to younger ground-state scientists like myself. However, I was not the only one who saw the halo, judging from the way the eyes of all attendees (not to mention the speaker) would turn to him whenever he quietly asked some questions or made some comments. Meeting him personally certainly refined, but ultimately increased my religious reverence, as witnessed by my little poem in Italian and my drawing below. Both were done right after coming back from a brunch at Mara and Walter's, where I had the very pleasant surprise of being invited together with a few other participants in the program. My art work was shown to some of the Italian colleagues there (Stefano Baroni, Roberto Car, Raffele Resta, Annabella Selloni, Gaetano Senatore) and has since been hanging in my office in Rome. It was not until last year that I dared give a copy to Walter.

## The Nobel Prize

The news that Walter had won the Nobel Prize in Chemistry reached me in China, where together with Mario Tosi and Norman March I was attending

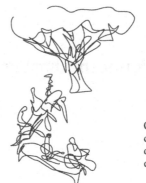

A colazione
con Walter Kohn
oh, ch'emozione
oh, ch'emozion!

Mangio insalata
fra i sapienton,
fra famosissimi
gran cervellon.

Ma se poi sbaglio
qualche question,
prendo un abbaglio
sull'elettron?

Oh che figura,
che umiliazion!
che pomodori
da Walter Kohn!

Drawing and poem after my first visit at Mara and Walter's home (and garden) in Santa Barbara, July 1994. The English translation reads: A brunch at Walter Kohn's / how exciting, how exciting! / I'm eating salad among wise guys / among most famous Big Brains. / But what if I miss some question / I make a blunder on the electron? / What a poor figure, what a humiliation! / Rotten tomatoes from Walter Kohn!

the 9th International Symposium on the Physics of Materials. I was very proud to add a handwritten line of breaking news near Walter's name in one of my first viewgraphs. Proud? Of course I was not the inventor of the Density Functional Theory – I was just one of the hundreds, maybe thousands of its users and developers in the field of condensed matter physics. And besides, when Michael Schlüter and Don Hamann had introduced me to its mysteries, that marvelous theory was already fifteen years old. But even so, I felt that a tiny fraction of that glory was bouncing back to me as my memories went back, again and again, to the brunch in the Kohns' delightful garden.

## The Holy Year

Back in Rome, I was invited by Umberto Grassano, a good friend of my thesis advisor Franco Bassani, to join the program committee of one of the many academic events scheduled for the Holy Year 2000. Though an active Church member and a physicist, I was not sure I was the right man for the job. But Umberto was very serene when he insisted that my presence was needed because he was sick and might not be able to follow the work through (as a matter of fact, he died in May 2000). I decided to join in and the committee agreed that, rather than having a conference of 'Catholic scientists', we would ask a top-level, religion-blind set of speakers to illustrate the 'Physics of the 21st century', with just one panel session devoted to religious and philosophical discussions. Each one of us was supposed to find speakers of appropriate standing within his or her own field of research. The first person that came to my mind was Walter Kohn, so I wrote him an e-mail message that Walter immediately answered.[1] His reply was very friendly: Walter told me that he had been raised in an ecumenical spirit and had many Catholic friends, but in his view the Vatican had missed the opportunity offered by the Jubilee to come to terms with its great failures during the Holocaust, some of which Walter had experienced firsthand (e.g. Vienna's Cardinal Innitzer greeting Hitler's envoy with the Heil Hitler salute). He reminded me that his parents and other dear ones had been murdered, and suggested putting together – and bringing to the participants' attention – a list of Jewish physicists who had been victims of the Holocaust, including his high-school teacher Emil Nohel, who had been Einstein's assistant. What could I reply? I admitted with shame that in the early stages of Fascism in Italy (and Nazism in Germany) the prevailing reaction of our highest authorities, but also of our local church leaders – and, alas, of ground-state Christians – had ranged from silent to enthusiastic. That certainly was not balanced by the few and soon exterminated Christian opposers of Nazism, such as the White Rose group in Munich, or by the Christians, nuns, priests and bishops who during the war had helped people chased by the Fascists and Nazis, as described for instance in Ben Cross' movie 'Assisi Underground', or as testified by the list of names displayed on the Avenue of the Righteous at the Yad va-Shem memorial in Jerusalem. I added that the Pope was expected to make a detailed, public apology for the historical responsibilities of the Catholic Church in early 2000. Finally I wrote Walter that in inviting him, besides obvious scientific considerations, I had hoped that our (bad) habits could be improved by a little help from 'good people' outside our Church. Again, Walter surprised me by replying almost immediately. And even more surprisingly, he wrote that one of his neighbors and friends in Santa Barbara was George Wittenstein of Hamburg, the only survivor of the White Rose group, whose name was totally unknown to me (all I knew was that all the group's members had been beheaded). Walter said he would like to wait for the Pope's statements on the mistakes of the Catholics before a final decision on his participation. He added that he, in his scientific talk, was going to include a brief remembrance of the Holocaust victims, particularly his own teacher; to mention the responsibilities of all those who had kept silent in spite of the moral authority they possessed; and to emphasize, for the future, individual and collective responsibility when human rights are violated or threatened. To make a long story short, in the end Walter came and gave a wonderful talk exactly in these terms. Rather than blaming anyone for failing to speak out, he asked the audience a constructive and much more poignant question: 'What are you going to do next time?' The conference was followed by a large papal audience in the Vatican, at the end of which Walter was one of the few who were admitted to see the Pope, while most of us watched them from a distance on a large TV screen.

---

[1] Another scientist whom I thought of inviting – and who was fortunately able to attend – was David Mermin; his contribution is included in this book.

I remember that his exchange with the Pope seemed to last longer than any other. Back to St. Peters Square, on our way to the hotel, David Mermin and I were quite curious to hear about it. In his own recollection, Walter's words had been something like, 'Your Holiness, I appreciated all your efforts towards a better understanding between the Christian and Jewish people', to which the Pope had smiled and replied with a repeated 'Thank you'. On the TV screen, that handshake had seemed to last forever. In remembering the time I had invited Walter, I was very touched by such a conclusion – but was that really a conclusion? On that very same day the Vatican document *Dominus Jesus* was published, which raised many comments both within and outside the Catholic Church. Walter in particular did not view it as a terribly promising new step in the direction of inter-religious relationships. Back in Santa Barbara, he felt the urge to express his concern directly to his new friend the Pope, so I gladly assisted him in the most unusual task of trying to have his personal letter delivered to the Holy Father. Monsignor Zycinsky, the Archbishop of Lublin and a Physics Ph.D., who had met Walter in Rome at the Jubilee of the Physicists[2] was of great help, so that a few weeks later, to my surprise, Walter told me that he had already received a (nice) answer from the Vatican. A longer account of this experience, in Walter's words, can be found in his talk, 'Reflections of a Physicist after an Encounter with the Vatican and Pope John Paul II,' given at UCSB in April 2001.[3]

## Last Summer in Santa Barbara

After that memorable time Walter and I kept in touch. Last summer, the ITP program on Realistic Theories of Correlated Electron Materials, headed by Ole Andersen, Antoine Georges and Gabi Kotliar, offered an opportunity to spend some more time together. When my turn came to give a talk, not only was Walter the first one to enter the room: he also threw me a life ring after a difficult question from Werner Weber. Apart from the official program, I learned a lot on new physics both at the Chemistry lunchtime seminar, to which he invited me, and in private conversations in his office, one of which was accompanied by Chinese tea of a very exclusive brand, prepared with his own hands. Another time, the Einstein pictures hanging in his office were the starting point for a moving and intellectually amazing account of the life of their author, a famous biologist and photographer who had first fled into Berlin from Russia after the Revolution, then from Berlin to the USA after the rise of Nazism. The photographer was none other than his father-in-law Roman Vishniac.

The rest of my family also had the privilege of Walter and Mara's company as they generously gave additional informal parties at their place besides the official ones. In such occasions Walter will deliberately leave physics out

---

[2] A contribution by Mons. Zycinsky is also to be found in this book.

[3] see http://www.srhe.ucsb.edu/lectures/info/kohn.html

and allow everybody to enjoy his company and his humorous way of sharing his amazingly vast and profound human experience. At our program's social dinner I remember Walter walking into the AppeThai restaurant, just by chance, among our four kids; his first joke was to tell the waiter that he was their grandfather – *il Nonno*, he explained to them in Italian. In the following weeks we enjoyed many more happy times with Mara and Walter. The guacamole prepared by Walter's hands was as delicious as the conversation. The hats and sunglasses offered to the guests made the hospitality perfect. My wife Silvia was surprised to see that the only visible sign of Walter's Nobel Prize in their home was a small picture[4] stuck with a magnet to the fridge among other family pictures and notes. In Silvia's view, that was very revealing of the Kohns' style and approach – it showed how proud Mara was of her husband, but also that their children and grandchildren were no less important to them than the King of Sweden. Even my children felt as important as kings when, on our last day in Santa Barbara, Mara paid us a surprise visit and brought them many farewell gifts including a few beautiful pictures of dolphins which are now hanging in their rooms at home.

Thank you, Mara; thank you and happy birthday, *Nonno* Walter! A spiritual and moral guide, a father and a grandfather – not just the Founding Father of the Density Functional Theory.

**About the Author**: Giovanni Bachelet is professor of Physics at the University La Sapienza of Rome. His research interests are in condensed-matter physics, and have covered the development of norm-conserving pseudopotentials within the Density Functional Theory, their application to a variety of solid-state systems and their extension to quantum Monte Carlo simulations. His most recent activity concerns the 3D and 2D electron gas and the electron-phonon interaction in magnesium diboride and related compounds. Contact: Dipartimento di Fisica, Universitá La Sapienza, piazzale Aldo Moro 2, I-00185 Roma, Italy; Giovanni.Bachelet@roma1.infn.it; axtnt2.phys.uniroma1.it/GBB/

---

[4] Similar to that appearing on the first page of this book

# Walter Kohn, the Chemist

Evert Jan Baerends

Vrije Universiteit, Amsterdam, The Netherlands

I met Walter Kohn for the first time after a talk I had given at a quantum chemistry conference. Although my whole life is being spent in trying to understand and apply density functional theory, I happened to be in a critical mood that day and was somewhat harshly criticizing the theory, or rather the local density approximation – it was before the GGAs made a breakthrough – for its rather unphysical assumptions when applied in molecules. Walter might have ignored me as the uneducated young upstart with unfinished ideas which I must have been in his eyes. It is very typical for Walter that instead he sought me out after the session was over, and addressed me without any sign of annoyance or any attempt at patronizing. We had a long and quite interesting discussion (to me) about both the foundations and the practical implementation of DFT. I felt great afterwards, very pleased with the fact that it proved possible to have such an open and lively discussion with someone who was the absolute top authority in the field. By focussing on "the heart of the matter" rather than wasting time on the question "who do you think you are, young man", Walter right away established himself for me as my role model in the world of science.

I have met Walter many more times since then, and heard him speak and ask questions on many occasions. By now I know that my experience was not quite typical for how life in the scientific world usually is. It was just very typical for Walter. The Dutch physicist Ad Lagendijk once coined the phrase "the arrogance of physics" to denote a tendency among physicists to value very highly the fundamental research into the most general laws of nature and to give a lower standing to those parts of physics that deal with the complexities (sometimes perplexities) of phenomena governed by known laws. The example of the latter type of physics would be chemistry. Sometimes this scientific value system translates into arrogance towards the more down to earth physicists that deal with manifestations of matter and radiation rather than with their origins. And could we deny that there are physicists who have a condescend attitude towards chemistry and the chemists?

Walter's contributions rank among the most fundamental ones, so fundamental that it has taken quite a while for their relevance to become fully clear and their practical scope to fully unfold. At the same time, Walter Kohn is the epitome of the non-arrogant physicist. This personal quality pervades, I think, his whole approach to scientific problems, and indeed I cannot help thinking that it helps to understand the nature of his scientific achievements. The lack of arrogance and prejudice made him never turn away from problems out of contempt for – seemingly – less lofty regions of the scientific endeavor. That directly leads to the well known qualities of his work of getting to the heart of the matter, uncovering the most general and simple and therefore most consequential relations.

Isn't it beautiful that the least arrogant of the "deep" physicists, was bestowed with the highest honor in the world of science by the chemists? And that is no accident. His unprejudiced approach, in this case to electronic structure theory, has led to the derivation of relations that are as important in chemistry as in physics. There is also an amusing side to this. Walter is a true physicist in his search for simplicity and generality. Chemists are not like that. They like pluriformity, the world of many different compounds with an infinite number of different characteristics. As if there aren't enough already, they continue to make more. The more the merrier. I have seen Walter listening to many chemical lectures and have wondered how bewildered he might have been by all these structures and great variety of properties and purposes. But always Walter showed his great appreciation for this new world of color and variety. Only once I saw him somewhat put off by a very direct question. After a talk on catalysis, with intricate orbital explanations of what was going on, the speaker suddenly turned to Walter and exclaimed: we are always using these Kohn–Sham orbitals as if they are just like our familiar (Hartree–Fock, extended-Hückel) orbitals, but what are they really? Is there any justification for what we are doing? Walter became very cautious. There is one thing we are sure of, he said, which is that they build, in principle, the exact total density. It is also clear that they look pretty decent, not so much different from the orbitals we are used to. But, he said, whether you can do all those difficult things with these orbitals that chemists tend to do, like arguing about $s$ bonds and $\frac{1}{4}$ bonds, donor-acceptor interactions and what not, I am not at all sure of. Be careful.

Here the great physicist proved to be a too cautious chemist. Since that time I have been preaching the gospel of the Kohn-Sham orbitals as tools for chemical molecular orbital analysis. They are better than anything we had before. They are even better than their originator thought they were. This begs the question: was their creation physical intuition or chemical intuition? Let us say that the Nobel committee has answered that question! We may frankly speak of the chemist Walter Kohn.

Evert Jan Baerends

**About the Author**: Evert Jan Baerends is professor of Theoretical Chemistry at the Vrije Universiteit, Amsterdam. His main scientific interest is density functional theory, both the theory itself, as well as the development of computational schemes based on DFT (the ADF program system), and the application of electronic structure theory to the understanding and prediction of chemistry. Areas of particular interest have been relativistic effects in heavy element compounds, time-dependent DFT, chemical reactivity and homogeneous catalysis, and surface-molecule interactions and heterogeneous catalysis. Contact: Department of Chemistry, Faculty of Exact Sciences, De Boelelaan 1083, NL-1081 HV Amsterdam, The Netherlands; baerends@chem.vu.nl

# Looking 50 Years Back:
# Walter, on the Occasion of His 80<sup>th</sup> Birthday

Sidney Borowitz

New York, U.S.A.

It doesn't exactly seem like yesterday but I was jolted when I was confronted by the fact that we first met more than fifty years ago – more than half a century. Whoever thought of celebrating your birthday with a volume of recollections of friends did everyone a favor. It should bring you great joy. For the contributors if affords them the opportunity to relive a part of their life with you in the background.

You will recall that we first laid eyes on each other in an office of the Jefferson Laboratory of Harvard University. I was to be an Instructor of Physics with duties prescribed by the department chair and was to look to Professor Schwinger for any research interests he would ask me to pursue. You were probably given the same instructions with the possible exception of having to look to Professor Schwinger for guidance for your research activities.

I don't believe I ever discussed with you what my feelings were when I first arrived at Harvard – I was overwhelmed. It was a milieu like no other I ever was in. I was told and understood that Harvard did not have to keep up with the Joneses, because they were the Joneses. The superb green lawns could have been the result of careful maintenance for three hundred years as far as I was concerned. And when I got to know the students who were studying there, and got to know you better, I realized how unprepared I was for the opportunity I had been offered to augment my education in Physics.

My arrival coincided as you know with one of the most productive and exciting moments in the history of modern physics, the ability to do quantum electrodynamics without having to worry about the infinities that had plagued previous researchers in the field. The role I was to play was to help clean up some of the problems left behind when previous calculations in this field could not yield results that could be compared with experiment. I managed in a relatively short while to be able to do calculations fairly quickly and efficiently. However it became clear to me that I could not master the subject well enough to have it serve as a basis for a research career. I came to realize that if I were to do physics it would have to be in some other subject.

I suspect that you came to the same conclusion about your future but with a difference. You started your future life by doing research in fields other than quantum electrodynamics immediately. I learned this because you naturally told me about some of the things you were doing.

I saw that my future at Harvard was all behind me and started looking for another position. I was offered an Assistant Professorship at New York University under circumstances that still give me an eerie feeling when I think of it. You will recall that I was made the offer shortly after a young recent Harvard graduate, whose name I do not remember, died of a heart attack before he could assume the same position I was offered.

At New York University I undertook research in the field of scattering in atomic systems. I believe that I performed adequately as a professor in teaching and thesis supervision. But after a few years I realized that being the research physicist I became did not satisfy me. I aimed to change my departmental duties. Without neglecting my professorial work, I volunteered to do administrative work for the department. This became an easy transition when Zumino became department chair. He was grateful for a helping hand because he had other fish to fry in his research activities. He rewarded me by naming me officially Assistant Chair. And when he left the University I was appointed Department Chair.

When a decanal position became vacant a search committee recommended me as Dean of the University College Dean of Arts and Science. Shortly thereafter the Chancellor of the University left and I was offered and accepted the position of Chancellor and Executive Vice President.

This saga started with my years at Harvard that were among the happiest in my life. I never completely lost the feeling during the two years I was there how great it was. It was the atmosphere, the people I met, and not least that I was able to find you as a friend. Sharing an office with you for two years set me to thinking that I should look for another career. It was pure luck that I was able to do so in a socially acceptable way. My years as a Chancellor convinced me that I did the right thing. I was a far superior academic administrator than I would have been as a physicist.

There are not many specific things that happened to me at Harvard worth recording. Perhaps you would not mind a few. Shortly after the semester began I was invited to attend a final oral exam for the Ph.D. After it was over Van Vleck approached me and put his arm around my shoulder and said, "Don't worry Sidney. They are not all as smart as Abe Klein".

I recall that we used to bring our lunch in a paper bag, furtively lock the office door and eat. One day we received a note from Van Vleck saying he wanted to have a departmental meeting in his office during lunch hour, suggesting that those who bring their lunch from home to bring it to the meeting as he does every day. At last we need not be ashamed, we were at least part of the mainstream if not of the elite.

The last recollection I will recall was a meeting of the physical society at which I presented a short paper in collaboration with Schwinger (and possibly you). As I passed Pauli, who was sitting on the aisle, I heard, "Hier kommt der Trabant (satellite)". I wonder if this remark set me thinking.

Well, happy birthday, best wishes and congratulations on your life's accomplishments.

**About the Author**: Sidney Borowitz is an emeritus Professor of Physics at New York University. Despite his title he is a lapsed physicist having gone into academic administration in 1970. His penultimate title at New York University was Chancellor. Before that he was a Provost of a campus that followed his having been Dean of an undergraduate College of Arts and Science. He merits inclusion in this volume because he shared an office with Walter for two years as an Instructor of Physics at Harvard University and mantaining a friendship with him to this day. Contact: 70 East 10th Street, New York, N.Y. 10003, U.S.A.; sb8@nyu.edu

# Meeting Walter Kohn

Ewa and Krzysztof Brocławik

Polish Academy of Sciences and Jagiellonian University, Cracow, Poland

Walter Kohn became the most important and meaningful person in my life and that of my husband Krzysztof. Our experience is based on different perspective in spite of the fact, that most of the events we are talking about took place in the same place and at the same time. That is why this story is written in two parts. In the first part "I" means Ewa Brocławik, theoretical chemist, while in the second part "I" means Krzysztof Brocławik, at that time lecturer at the Faculty of Psychology, Jagiellonian University.

## Ewa's Story

I met Walter Kohn for the first time in Il Ciocco in 1993 at a NATO Advanced Study Institute devoted to density functional theory. I was then at the beginning of my scientific way but had already first experiences with DFT, which had then just entered the chemists' world as means to probe chemical reality at atomic scale. This conference was predominantly a meeting of physicists and Walter obviously was the member of honor on the board. I was the chemist in the sea of physicists. Nevertheless, my deep feeling persisting from that time is that the dialogue between the two societies, the physical and the chemical, was already very much alive then and became more and more vivid ever since. From the first minute of our first contact I noticed that Walter presented the type of an actively listening personality, his glaze was always full of goodwill and apprehension. He would listen with high interest and, at the same time, tremendous peace of mind so I never felt to be lower in rank as a chemist presumably asking ill-defined questions or using vague and not clearly-cut definitions.

Together with the Syrian Ihsan Boustani, whom I had met and made friends during my scientific stay in Berlin, we used to spend quite a lot of time with Walter during coffee breaks, meals and leisure time. We vividly discussed mostly culture, history and rudiments of life. Walter also told stories from his family life. I was completely taken by his friendliness, worldly wisdom and this very special "chemistry" of open attitude towards different cultures, religions and life attitudes, and values, which gave to our somehow strange company an atmosphere of close relationship. I feel emotionally, that not only his unquestionable position in science but also just his personality, peace of mind, goodwill and wide interests made him such an exceptional and inspiring person.

I recall, that it was not a big surprise for me when the next year I received a personal letter from him: In 1994 the international symposium "Thirty Years of DFT" was organized in Kraków. Walter was of course invited to the conference as the honorary speaker. He asked me for the favor to be his personal guide in Kraków, where he wanted to spend some extra few days before the conference in order to trace down some memories from his family life. He wished to visit privately the city including the Jewish quarter called Kazimierz and to make the trip to Auschwitz. I was of course only extremely happy to meet his request.

It happened that we again met Ihsan Boustani, my Syrian friend, and Lu Sham and his wife. Together we visited Kazimierz, already renovated in large part. Due to the personal attitude of Walter and his wise stories this was a very important experience also for me. Of course, being a thirty-year inhabitant of Kraków I had visited Kazimierz many times before. However, this time I experienced quite a different perspective, understanding, and insight.

A very special event appeared to be the photograph exhibition "Cracow Jews 1869–1939" in the Old Synagogue, most of them by the famous photographer, Roman Vishniac, who, if I recall it well, was the father-in-law of Walter.

The second wish of Walter, the visit to Auschwitz was assigned to Krzysztof and his story is as follows.

## Krzysztof's Story

For many years I had a special kind of unofficial mission. I served as a guide to Auschwitz for many honorable scientists, who visited the Faculty of Psychology at the Jagiellonian University. Auschwitz has already been known as the horrible place, however, many people want or feel obliged to visit it. I was then not surprised when Ewa asked me to guide Walter there. On the agreed day three other attendees of the conference wanted to go to Auschwitz as well. I have to confess that, according to my experience, people tend to react to Auschwitz in very different ways. I am not going to value these reactions and ways of experience, however, I must stress again that being there is a very tough and heavily emotional experience.

This time I had in my car Walter who, what was very well known to me, was personally affected by the Holocaust. I also had a German, a Syrian and a Canadian of British origin. The atmosphere in my car was thus from the very beginning a mixture of hanging out tensions. Neither serious themes

A photo taken in 1994 at the cafeteria in Kazimierz quarter of Kraków. Sitting by the table around me is (to the left): Ihsan Boustani, Walter, Lu Sham and his wife (photo: E. Brocławik)

would be developed nor jokes and stories would be listened to. The day was exceptionally beautiful. When we arrived at Auschwitz Walter wanted to see the entire place in detail and the other members of company agreed to do the same, maybe driven by the respect to Walter. However, when we were walking through the camp, one by one our friends gave up and disappeared. Very quickly we, Walter and me were alone on this route.

On our way we came to the exhibition room with several hundreds of documents concerning camp life and affairs displayed. I got the impression that Walter just briefly glanced around. Apparently this was not true as Walter suddenly pointed out to me a sheet of paper not bigger than a postcard. The content of this piece of paper, however, expressed in German bureaucratic slang a very important message about human fate in the camp. The same situation was repeated several times on our way. Walter glanced at a wall full of documents and precisely pointed out a very important or characteristic detail. On that morning I recognized and understood clearly what kind of minds are able to make important discoveries in science.

In this mood we completed the whole route across the camp. Our friends waited for us patiently in the near-by coffee shop. On our way back nobody gave any comments. On our farewell Walter advised me to read the book by Primo Levi, an Italian Jew and also a scientist, who survived Auschwitz. I was very well familiar with the written documentation about the camp but I did not know this book. After some time, I cannot recall exactly which way,

I received a copy of this book from Walter. I keep it in my library and regard it as one of the best personal witnesses about Auschwitz.

Walter will soon turn eighty! What would this mean for his friends in life and in science, to people whom he has influenced so deeply just by meeting them, talking to them or merely being present in close vicinity?

**About the Author**: Ewa Broclawik is a professor at the Institute of Catalysis and Surface Chemistry, Polish Academy of Sciences in Cracow. She cooperates closely with the Department of Theoretical Chemistry, Jagiellonian University. Her scientific interests concern quantum theoretical research on catalytic active sites. Her current activities include applications of advanced first-principles methods (including density-functional theory) for molecular modeling to elucidate mechanisms of catalytic reactions on surfaces and in enzymatic active sites. Contact: Institute of Catalysis and Surface Chemistry, Polish Academy of Sciences, Niezapominajek 8, PL-30-239 Cracow, Poland; broclawi@chemia.uj.edu.pl; atom.ik-pan.krakow.pl

# Of Sandboxes and Silences

Kieron Burke

Rutgers University, Piscataway, U.S.A.

Walter Kohn has been an enormous influence on my professional life. I was one of his later graduate students (PhD, 1989) and managed to create a home for myself in chemistry, by working in density functional theory. A pinnacle of my professional life was, therefore, to spend eight weeks last summer visiting him in Santa Barbara.

With their customary kindness, so well documented in this book, Walter and Mara invited us (my family, including three small children) for a barbecue at Kohn ranch. Fear swept over myself and my wife, Sharon, remembering all the beautiful and valuable objects arrayed around that house, and imagining the utter destruction our rambunctious children might cause. We had no idea how to keep them out of trouble for any length of time, and devised emergency plans A, B, and C, for a hasty departure, if need be.

When we arrived, we discovered a variety of distinguished guests, none of whom had brought even a single child. But Mara, calm and confident, led us down the winding garden path where, in a corner just before the end, she asked me to pull back a tarpaulin, to reveal a shaded sandbox, with a variety of plastic toys. The children were delighted, and essentially vanished until food was ready. Mara's thoughtfulness meant that we actually enjoyed our visit. Later, Walter sat at the children's table, and even persuaded my eldest son to partake of his barbecued chicken, a feat I have since been unable to duplicate (must be the recipe). How many of you reading this book have walked down that garden path, but never noticed that sandbox?

Professionally, Walter has also sown a vast and exotic garden. Many of the plants within it are now well-known and recognized. But I still find, often in the shade, long forgotten sandboxes, containing buried treasures that bring more pleasure than many of the more famous works.

I have been lucky enough to discuss many scientific questions with Walter during my career. During the summer, we often had lunch together, sitting at one of his favorite spots at UCSB, on the bluffs overlooking the Pacific Ocean. One day, I raised a question I'd been working on with a graduate student for about a year. Before I could tell Walter what we had done, he indicated that I should shut up, which I promptly did. About ten minutes passes, during which we both munch our sandwiches in the beautiful sunshine, nary a word being spoken. Finally, he tells me where we should start, mentioning a paper by Peierls from the 30's. I say that we did that, and tell him what we found. Ten more minutes of silence, while we walk. Then, another calculation is suggested, which we had done, and I tell him the result. A last round of silence, followed by his last suggestion, which anticipates the work we were just completing. In about half an hour, he had recreated in his head our entire year's work!

My point is this. Walter has always emphasized the importance, not of finding answers, but of asking the right questions. When one asks him a well-considered question, the most wonderful response (to me) is silence. This indicates both that the question is interesting *and* that he does not immediately know the answer, despite his immense experience and intuition. Then you know you have a question worth answering!

The applications of Walter's work to biochemistry are just beginning. More and more density functional papers are appearing in which large biomolecules are treated. A friend is doing hemoglobin, I'm working on hydrogen bonds in protein-folding models, etc. I believe the best is yet to come, and that the 100th birthday celebrations will be with an even wider circle of friends and admirers. May they also learn from silence, and their children enjoy that sandbox.

But for now, happy 80th birthday, Walter. I thank you and Mara for everything. We chemists truly appreciate you.

**About the Author**: Kieron Burke is a Professor in the department of Chemistry and Chemical Biology of Rutgers University, and a member of its graduate program in Physics. His scientific interests include quantum chemistry, materials science, condensed matter theory, biochemistry, and surface science. Contact: Dept. of Chemistry & Chem. Biology, Rutgers University, 610 Taylor Rd, Piscataway, NJ 08854, U.S.A.; kieron@dft.rutgers.edu; dft.rutgers.edu

# A Glimpse of Walter Kohn

Sudip Chakravarty

University of California, Los Angeles, U.S.A.

We finally made it to San Diego, all across the country and the stark expanse of the southwest, in a red Volkswagen Rabbit with no air conditioning, my wife, a pet dog, a myriad potted plants, boxes of photocopied papers, stacks of computer outputs and not much else. I was going to be Walter Kohn's postdoc. All throughout the trip I thought about what it was going to be like. Vivid accounts of Walter from my mentors at Cornell made me almost certain of his looks, his worldview, and his sharp critical mind. I felt that I had known the person since my childhood, although I had never seen him or met him. There was always a Walter story floating around at Cornell, thanks to Vinay and David, respectively a former student and a postdoc. In short, I came all prepared, as though this was my destiny. My wife, Nancy, who is not a physicist, and had so far met only the fun-loving, quirky physics types, did not have a clue of what was going through my mind, as I am sure went through the minds of many past and future wannabe postdocs of Walter.

I finally met him in his office with a boyish mischievous smile on his face, entirely shattering this great fantasy of mine. He seemed utterly pleasant at lunch, expressed his frustration at not quite being able to formulate a dynamical version of the density functional theory, more intrigued than tormented. I knew right away that I would get along very well with this man, who is just like my other fun-loving, quirky friends. So, what was all this fuss about Walter? I gradually started getting a glimpse of it as I discussed my interest in statistical mechanics with him – precision of thought was the key. I was working on a utterly useless problem of proving that despite long range interaction in a one-component Coulomb plasma, Mermin's proof of the non-existence of crystalline order in two dimensions holds. I was very pleasantly surprised that he was so encouraging in spite of the fact that this was not what he was interested in at that time. "Everyone needs to prove a result in mathematical physics once in his lifetime", he said. "I once *proved* that Wannier functions are localized in one dimension, a piece of nineteenth century mathematics, but a proof it is. I pull out a Wannier function out of my desk drawer, when people start talking too much." He was looking very happy those days, and later I found out why. His yearly trip to Paris was coming up, something that he dearly loved. So, Martin Fogel and I, another former Cornellian postdoc with Walter, scrambled around to do something about the dynamics of density functional theory, and we had great fun, and in less than a couple of months we had a paper in the revered journal called the Physical Review Letters. In retrospect, I am not quite sure what we proved there, but it did not divert us from the joyousness of the event. Do not ask us what it was about, because the statute of limitations is passed. We are not to be held responsible for it. Walter took off happily for Paris, and I was more or less left to my own devices, nominally under the supervision of Shang-keng Ma and Harry Suhl, two great physicists at San Diego.

Walter returned from Paris and informed me that he was going to be the Director at the then Institute for Theoretical Physics in Santa Barbara, and wanted to know if I would be interested in following him. This was disclosed, if I remember it correctly, at a New Year's Eve party that Walter threw at his Del Mar house, for which he had prepared the most wicked punch I have ever had. He was clearly pleased with the effect it produced in us. Actually, I am not sure if he himself had any of it. He was very enthusiastic about his new job and correctly sensed that he would have a hand in shaping the agenda of theoretical physics for years to come. Nancy and I, and our trusty hound, tagged along. The Institute became a unique place for me. I never learned as much from any other place. In spite of his directorship, Walter and I discussed physics everyday, never agreeing on anything. But I saw his mind in action. The problems that we discussed there haunt me to this day. We discussed what are Mott insulators and localization, new approaches to real space quantum renormalization group, gauge principles, and implications of lattice gauge theory (An excellent program on lattice gauge theory was in session at that time at the Institute.). I studied a then-forgotten but profound paper by Walter on the characterization of Mott insulators. I was struck by it and Walter helped me with it. We discussed topics that were far ahead of the time. My only regret is that I wasn't able to grab onto them in a tangible manner, but we enjoyed our daily dose of arguments. Another dear friend of mine, the late Albert Schmid, who was visiting the Institute, wrote a few pages on these daily Chakravarty–Kohn skirmishes in his parting report, which Walter gleefully showed to me. In the midst of exciting physics, Walter paid meticulous attention to the details of the Institute. In fact, he knew practically every postdoc's long distance phone bills! He had a running battle with the maintenance guy who used to fix the elevators in Ellison Hall. Once, about ten physicists, including Walter, were stuck in the elevator for a considerable amount of time. When they finally got out, Walter belted out to the maintenance guy "Can you imagine ten distinguished physicists stuck in your elevator?" , only to elicit the response: "Poor elevator". Evil humanities types who used to inhabit the lesser floors of the Ellison Hall used to whisper that Walter had rigged up the elevator in such a way that it went straight between the first floor and the sixth floor and only whimsically stopped in between.

My wife and I were invited to Walter and Mara's house quite frequently and so were the many Institute postdocs and visitors. It felt like a family. Mara was simply wonderful, with her hippopotamus Hubert next to the fireplace. She was a source of comfort at difficult times and always full of energy and advice. But she sometimes misjudged Walter, as I remember her throwing a surprise birthday party for Walter. We were lucky that we were not the first to arrive at their house, a fate the poor Cornwalls suffered. We loved Mara immensely and showed her all our accomplishments, however little they were, and she was always ready to cheer us. I remember our daughter met her for the first time when she was just a few years old, and she loved her, not because she was an overwhelmingly grandmother sort, but she respected her individuality, even when she said "no" to something that was suggested to her. I also remember them aimlessly wandering around together in their beautiful garden one sunny afternoon, as Walter and I discussed physics, and Nancy read a New Yorker article in their patio. It is not just us, but whoever came in contact with them felt the same way.

There are so many memories to write about, but alas Marcel Proust has already done it, and I could not be possibly immortalized by my writing. But there is one aspect that I am sure no one has written about. Walter actually played flute rather well, is a great tenor and knows Christmas songs by heart. I still remember a holiday party at the Institute, where Steve Kivelson, Walter and I sang "Joy to the World". Now, imagine that! I distinctly remember walking into Walter and Mara's house in Santa Barbara as I heard the sound of one of the Händel's Hallenser sonatas. It was Walter playing, without the slightest bit of practice, on his french open-hole style flute. In his characteristic style, he said "Try it out yourself". It is like him saying "Hey, I know you can do that integral". I picked up the flute, made a perfectly respectable sound out of it, and I have been playing flute for the last twenty five years. I remember when I first learned to play the Bach E-minor flute sonata, Nancy and I made a special trip to Santa Barbara to play for him. There are very few in this world who are capable of communicating their gifts to others so effortlessly and generously. Walter is one of them.

**About the Author**: Sudip Chakravarty is a professor of physics at the University of California Los Angeles. He is interested in high temperature superconductivity, correlated disordered systems, and in dissipative quantum mechanics on a macroscopic scale. Contact: Dept. of Physics & Astronomy, UCLA, P.O. Box 951547, Los Angeles, CA 90095-1547, U.S.A.; sudip@physics.ucla.edu; www.physics.ucla.edu/people/faculty_members/chakravarty.html

# Quantum Reflections

Dennis P. Clougherty

University of Vermont, Burlington, U.S.A.

When I was a student at MIT, Victor Weisskopf often spoke about the privilege of being a physicist. Physicists are provided with resources to conduct research on problems that interest *them*, and only a report of the findings is asked for in return. No demand is made for a service or product that succeeds in the marketplace in producing a profit. Yet, physicists enjoy a comfortable living, ample opportunity to travel, constant intellectual stimulation, and an opportunity to influence those who follow. In 1989, I was doubly privileged because I was invited to postdoc with Walter Kohn.

Joining Walter's group is a bit like being drafted by the Boston Celtics or the New York Yankees. You can not help reflecting on the past champions that wore the uniform. You feel great pride with the association. You are grateful for the chance to develop your skills with talented team members and a famous coach. And you hope you have a chance to hit one out in the big game, adding to the reputation of the storied franchise.

I fondly recall our meetings in his corner office at the top of Broida Hall. As one entered the office, there was a rectangular table in the middle of the room. There were filled shelves to the left and under the window sill to the right. There was a seldom-used computer on a table in the opposite corner. The exterior walls were large windows offering breathtaking views of the Pacific. Behind the door was the blackboard where his students and postdocs attempted to defend themselves with a small piece of chalk. It was rarely sufficient against the onslaught of questions. On occasion, Walter could get you to doubt the rules for multiplication.

Directly across from my usual position at the meeting table were three framed items: a photograph of Walter receiving the Medal of Science from President Reagan; a photograph taken by Roman Vishniac (Mara's father) of Einstein contemplating a set of field equations on a blackboard at Princeton; and a colorful scroll from the ITP given in appreciation of Walter's service as its first Director. I always found these things inspirational.

There was also a framed message expressing something of a Buddhist sentiment; it said roughly that instead of dwelling on things one wants but doesn't get in life, one should think of all the things that one doesn't want and doesn't get. It's turned out to be very good general advice for living. Working for Walter was always far more than an apprenticeship in theoretical physics.

On Arroyo Burro beach in Santa Barbara after a walk in the summer of 1998

When I first arrived in Santa Barbara, I had anticipated that we would work on something related to density functional theory. Hardy Großwas a Heisenberg Fellow at UCSB then, and he and Walter were involved in extending DFT to calculate excited state properties. Walter however had another problem in mind involving low-energy atom-surface collisions: what is the asymptotic form for the probability of an atom to stick to a surface as the incident energy of atom tends to zero?

The problem was discussed in the early days of quantum mechanics by Lennard-Jones using perturbation theory. The perturbation result is amusing because it is maximally non-classical. For a neutral classical particle moving near an insulating target, energy is always transferred from the particle to the target in the process of the particle doing work in mechanically deforming the target. This energy is radiated away by phonons, never to return to the particle. As a result, a particle with infinitesimal energy cannot escape the region near the target after the energy loss, and the particle is stuck to the target. The probability of a classical particle sticking is unity.

Lennard-Jones discovered that, for a quantum particle, simple wave mechanics reduces the likelihood of the particle's probability density near the target in proportion to the energy. This effect is now referred to as "quantum reflection." The result is that as the energy tends to zero, the particle spends less time in an area where it can transfer energy to the target. In perturbation theory, a quantum particle has zero probability of sticking in the limit of vanishing energy, in dramatic contrast to the classical case.

This quantum regime started to become experimentally accessible in the 1980's. Early experiments did not see the drop off in sticking with decreasing energy. A theory was proposed that perturbation theory was inapplicable in these experiments and that the strong coupling between the particle and surface deformations of the target led to a polaron-like collapse of the particle's wavefunction at the target's surface, overcoming quantum reflection. Walter found the proposed theory suspect and thought that there was work to be done here.

The problem appealed to Walter on many fronts; it involved quantum scattering, surface physics, many-body physics, analytic properties of the scattering amplitudes – all long-running interests in Walter's career work. Scattering theory has always held a special place in Walter's heart, and everyone knows the many important contributions he has made to it over his career. During our conversations, he often mentioned to me his long-time friend and collaborator Res Jost. Our work on quantum sticking took advantage of properties of the Jost function, and as a result I felt a connection through Walter to this historical thread.

We studied some theoretical models of sticking and concluded that the quantum reflection effect would eventually dominate at sufficiently low temperatures for a neutral particle striking insulating surfaces. The many-body effects simply postpone the inevitable vanishing of the sticking probability as the energy of the particle tends to zero.

Our conclusion was controversial at the time. Experiments then saw no sign of the downturn in sticking at the lowest accessible energies. The reigning theory was that quantum reflection could be overcome at sufficiently strong coupling. We submitted our results to Physical Review Letters. I probably have the unique distinction of having a paper (cond-mat/9205004), coauthored with Walter Kohn, rejected for publication. Undeterred, Walter suggested we submit a long detailed paper to Physical Review B; that paper was accepted in time.

Walter reminded me that we had gone out on a limb, predicting the threshold laws for neutral particles and for charged particles striking liquid Helium. It was obvious that lower energies would be accessible to experiment in the near future. Our results would be subject to direct experimental test. Fortunately two years later an experiment by Dan Kleppner's group at MIT saw the behavior consistent with quantum reflection in the low-energy sticking probability. It was a great relief.

Since that time, Walter has included me in some of his adventures. I finally had a chance to collaborate with Walter on a DFT-related project: inclusion of van der Waals interactions using time-dependent DFT. As part of that

Dennis P. Clougherty

collaboration, I met with Walter while he was spending the fall of 1999 at
ETH Zürich. Walter of course made certain that I took in the sights and
culture. We found time for a memorable walking tour of city, expertly guided
by Walter. He also introduced me to a drink, popular in Switzerland, called
Cynar – an artichoke-based liqueur whose appeal mystified me at the time. It
must be an acquired taste.

Those days at UCSB were special for me. There was a continuous stream of
interesting visitors, drawn to campus by a stellar faculty. Often Walter invited
these visitors to his famed brown bag lunch meetings – often emulated, never
fully duplicated without Walter's presence to keep the participants honest. I
also fondly recall the afternoon walks. We would discuss everything – physics,
art, music, travel, politics, history, technology – all on the beautiful beachfront
campus.

I will always be grateful for the opportunity to work with Walter as a
postdoc, and I am equally grateful for his friendship and advice over the
subsequent years.

**About the Author**: Dennis Clougherty is an associate professor of physics at
the University of Vermont where the large body of water to the west is Lake Cham-
plain. His research interests include topics in condensed matter theory and materials
physics. He is currently exploring the consequences of vibronic interactions in nan-
otube and fullerene-based systems, in addition to work on van der Waals interactions
in DFT. Contact: Department of Physics, University of Vermont, Burlington, VT
05405-0125, U.S.A.; dpc@physics.uvm.edu; www.uvm.edu/~dcloughe/

# Not a Relative, but ...

Marvin L. Cohen

University of California at Berkeley, U.S.A.

On September 11, 2001, I was in San Lorenzo de Escorial, Spain along with Walter Kohn, Matthias Scheffler, and perhaps others who are contributing to this book. We were attending the Ninth Annual International Conference on Density Functional Theory. As I left the auditorium after one of the talks that afternoon, I was approached by Steven Richardson who was a former postdoctoral from our group. As I stood speechless, he recounted the terrible events that had occurred in New York and Washington. I rushed to my hotel and could only get the news in Spanish. Although the pictures told most of the story, it wasn't until my wife and I joined others at the conference to watch the BBC in English that we learned the horrible details.

The next morning Walter was scheduled to talk. He faced an audience of agitated, concerned, and depressed people. We obviously were not in the mood to hear about Density Functional Theory, but that's what we were there for, and we felt a need to be together. Walter took the stage, and in a calm and thoughtful manner began discussing what had happened in the United States. Those of us who knew Walter were accustomed to his gift of going directly to the important, central issue of a problem in science, but I was still amazed at how he did exactly that in the midst of the emotional turmoil we were all experiencing.

Walter focused on the central issue-man's inhumanity to man. He'd seen it before. He recognized that this was a major crime against the innocent, and he knew what to say. Those of us who knew of Walter's past experiences were possibly even more touched by his calmness and his ability to find the words that calmed everyone else. His talk was appropriate and inspiring.

As we sat learning, understanding, and acknowledging that Walter had just said the things that should be said, Walter paused briefly and then began his technical talk. It was a signal that life goes on.

I was very impressed by Walter's speech at Escorial. All of us who spoke after him expressed our sadness and, although there were many eloquent words spoken, none had the impact that Walter's did. Although I've seen Walter since then, our Escorial encounter is the one I remember most vividly.

I do remember my first encounter with Walter Kohn. I was a graduate student in the early 1960's and I read his review chapter in Volume 5 of *Solid State Physics*, "Shallow Impurity States in Silicon and Germanium," written in 1957. I shouldn't say *read* – I practically memorized it. I looked up everything I didn't know or couldn't understand in the chapter. Since I didn't know much, I looked up a great deal of material. The chapter was so well written and clear, it directed my study of the material. I learned an enormous amount from this man I was yet to meet.

I can't remember my first face-to-face meeting with Walter. It was probably around 1963 or 1964. I was well aware of his work through many conversations with Quin Luttinger at Bell Labs in 1963-64. I studied the Hohenberg–Kohn–Sham approach as it came out and examined every argument. Although I trusted the results, I couldn't wait to test the method against experimental data. However, by the time I was in a position to use the DFT approaches, we had already solved the electronic structure problem for dozens of semiconductors using an empirical pseudopotential method (EPM). We were able to explain optical and photoemission in terms of interband transitions and had won every debate with experimentalists on the interpretation of their spectra.

I never had Walter's patience. His approach was to get all the fundamentals nailed down and make sure everything grew from there. I wanted answers to the experimental puzzles, and I wanted to make predictions. So I was happy with the EPM. In addition to their use in getting the physics right, they pointed the way for the *ab initio* theory. Of course, I was still strongly motivated to produce a method which required no experimental input. Along the way to first principles, we used semi-empirical approaches ranging from Fermi-Thomas, Wigner, and Slater potentials, but I was determined to employ DFT with the pseudopotentials. This became possible once a student John Walter and I produced electron charge densities. The Walter-Cohen paper produced some name confusion and considerable controversy.

Although chemists and physicists had visualized what the covalent bond in semiconductors like silicon and germanium would look like, many didn't like our pictures. The attacks by prominent chemists and physicists were harsh. In addition, it was at this point that I witnessed how much the chemists were suspicious and outright hostile toward DFT. It is ironic that Walter's Nobel Prize is in chemistry. It is also a lesson on how it sometimes takes a new generation to clean out the unfounded convictions of their older colleagues.

We implemented the DFT and converted our pseudopotentials to *ab initio* forms. At first I felt Walter was skeptical. I think it was because the methods grew out of empirical approaches. I remember after a seminar I gave showing that the electron densities were consistent with experiment and that we could predict and explain "everything" about silicon, he asked what experimental input I had used. I said, "nothing." He said, "really?" I said, "a little." I needed to know the candidate crystal structures to test, and when predicting superconductivity in compressed silicon, I needed to input the appropriate

Coulomb repulsion parameter. Walter's eyes twinkled, and he nodded making me very happy.

Finally the chemists joined the DFT bandwagon. In many of their calculations, they needed more precision than we did – so called "chemical accuracy." But on the whole they began to explore and use the DFT. Linus Pauling's reactions to our work were surprising. He was a great fan of our paper which calculated the total energy and structural properties of semiconductors. When I explained that the method was based on plane waves having no bonding prejudices, pseudopotentials, and DFT, he asked questions which showed that he understood what was going on. However, he referred to it as "your quantum mechanical approach" and said it wouldn't work for metals if I counted valence electrons in the same way as I did for semiconductors. I disagreed. He challenged me to do calculations for tin, but when I showed him the successful results, he was upset and then made a joke about physicists.

I told Walter that Linus believed DFT for semiconductors but he needed convincing about metals. We decided that Linus had a point concerning strongly correlated systems but he'd have to yield on itinerate electron metals. Of course by that time, the chemists were on board.

For many years I have given talks where I begin with Dirac's 1929 quote about quantum mechanics: "The underlying physical laws necessary for a large part of physics and the whole of chemistry are thus completely known, and the difficulty is only that the exact applications of these laws lead to equations much too complicated to be soluble." I refer to this as "Dirac's Challenge." Walter Kohn's contributions have helped us answer this challenge. I was therefore delighted when many of the public announcements heralding Kohn and Pople's Nobel Prize used this quote. Dirac would have also been pleased since he did research on DFT in the 1930's.

Let me close by relating an incident that occurred in my home when I was showing some guests a photograph of Einstein my wife had given me. I pointed out that it was a Roman Vishniac photograph and that Vishniac's daughter was married to Walter Kohn. One of the guests asked, "And how are you related to Kohn?" I answered, "Unfortunately, only through physics."

**About the Author**: Marvin Cohen is University Professor in the Physics Department of the University of California at Berkeley. He is presently engaged in theoretical research on electronic properties of solids, semiconductors, superconductors, and nanoscience. In 2002, he was awarded the National Medal of Science by President Bush and was elected Vice-President of the American Physical Society for 2003, President-Elect for 2004, and President for 2005. Contact: Department of Physics, University of California, Berkeley, CA 947200-7300, U.S.A.; mlcohen@uclink.berkeley.edu

# Riding Walter's Wake

Morrel H. Cohen

Rutgers University, Piscataway, U.S.A.

Young theoretical physicists or chemists nowadays cannot escape Walter Kohn's influence. Density-functional theory, created by an almost mythical figure occasionally glimpsed by the lucky ones, is everywhere. Those of us roughly contemporaneous with Walter have been even more strongly affected. During our decades-long struggle to understand more deeply the electron theory of atoms, molecules and condensed matter and to convert that understanding into quantitative descriptive tools with predictive power, we found over and over again that it was Walter who led the way, lifting the veils and shaping our subsequent thought. It was certainly true in my case.

The first post World War II generation of distinguished condensed-matter theorists included P.W. Anderson, J.M. Luttinger, and David Pines in addition to Walter. My own generation followed soon after, including E. Abrahams, R. Landauer, and A.W. Overhauser. We were aware of and influenced by the work of that slightly more senior generation. My first significant contact with Walter's work in particular was with the KKR method, which I studied though never actually used. Norman Rostoker, the R of KKR, left Carnegie Tech. for the Illinois Institute of Technology in Chicago somewhat after I arrived at the University of Chicago in 1952. His younger brother was in a class I taught in 1953, and through him I met Norm and learned something about Walter.

During the 1950's, Walter and Quin Luttinger spent summers working together at the Bell Labs, generating a remarkable series of papers on the physics of semiconductors, inter alia. One I paid particular attention to was their study of the g-factors of holes and electrons in semiconductors. I had been sensitized to the issues involved by my exposure at Chicago to Ed Adams' thoughts about crystal-representation theory, built on earlier work by Slater and by Luttinger. I was also at that time carrying out a joint research program at Chicago with A.W. Lawson on the group V elements and their alloys. It was thus natural for me to ask what the g-factor of carriers in bismuth would be. Soon thereafter, while on leave at the Cavendish Laboratory in Cambridge in 1957, I realized that the spin and cyclotron splittings were equal for electrons in Bi, giving a g-factor with components as large as 200 and explaining the puzzling alterations of sign in the harmonics of the deHaas van Alphen oscillations found by David Shoenberg.

Walter with his student Terje Kjeldaas and, independently, I with my student Michael Harrison and with Walter Harrison, then at GE, were all trying to understand the remarkable oscillations with magnetic field of the ultrasonic attenuation of metals observed first by Boml. Brian Pippard had constructed a simple geometric argument for transverse waves only, but the effect was seen for longitudinal waves as well. My colleagues and I worked out the theory for impure jellium and were able to capture all of the features of the experiments. Fred Keffer, a fellow Kittel student at Berkeley then on the faculty at the University of Pittsburgh, invited me to give a seminar on the subject at Pitt before we had published our results. Walter, then at Carnegie Tech., and Ted Holstein, then at Pitt, were in the audience. In the talk I went on and on about the formulation of transport theory needed for the theoretical analysis. Finally, Walter and especially Ted, losing patience with those irrelevancies, burst out with "But, did you find the oscillations?" Of course my answer was yes, but I could sense the unspoken remark "Well, why didn't you tell us that instead?" So, I was taught then by Walter to get to the point, to focus on the essential issues.

A year or so later, my first Ph.D. student, J.C. Phillips, while a post-doc with Kittel at Berkeley, initiated with Kittel's student L. Kleinman the modern theory of pseudopotentials. They first used it to obtain the band structure of diamond, and Jim used it to explain why the point-ion approximation worked in ligand-field theory. When I got their preprint, my instant reaction was that with the Landau Fermi-liquid theory and with pseudopotentials, we now understand why the free-electron model of metals works so extraordinarily well for the simple metals and their alloys. So I started giving talks all over the place, sharing my excitement. I was once again invited to speak in Pittsburgh, this time at the Mellon Institute (before the Carnegie-Mellon merger) by T.B. Massalski, a former Chicago postdoc. Walter, still at Carnegie, was again in the audience, sitting near the back. I was going on and on about how beautiful all this was when Walter raised his hand and objected that the Phillips-Kleinman pseudopotential was not unique because the overlap of the pseudo wave function with the core orbitals was undetermined. An analogous problem arises with OPW's of which I was well aware, but I had put off thinking about the issue, focusing instead on the marvelous interpretive power pseudo-potential theory provides. So, caught, I had to improvise, and, stimulated by Walter's question, I invented the pseudopotential cancellation theorem on the spot. I argued that the indeterminacy was a great advantage, as the overlap with the core functions could be adjusted to give the maximum cancellation of the screened ion-core potential. Apparently, I hadn't learned my lesson well enough at the acoustic attenuation talk. Once again, I was taught by Walter to get the essentials straight first, to complete the theory and then examine its consequences. I have delighted in telling this story over the years, but never before to Walter.

More than thirty years ago, Robert Pick spent the academic year at Chicago. Together with Richard Martin, then my student, we worked on the microscopic derivation of the Born-von Karman interatomic force constants, astonishingly not yet in the literature. At a certain point we needed to understand the wave-vector dependence of the matrix elements of the Coulomb interaction between the ground and excited states. Where else to turn but to Walter, who, with Vinay Ambegaokar, had worked out much of what we needed and provided the insight needed to get the rest. Consequently, we were able to derive the force constants, to define effective charges and to prove the acoustic sum rule.

Now, more than 40 years after these formative experiences began, I still find myself riding in Walter's wake. There are many examples of which I shall cite three. First, from 1981 through 1996, I was at the Exxon Corporate Research Laboratories. When I arrived in 1981, the laboratories were strong in electronic structure theory, and the theorists were able to collaborate effectively with the experimentalists in surface science, catalysis, and cluster formation, as well as in such areas of condensed matter physics as amorphous semiconductors. As a consequence of massive corporate-wide cuts and policy changes imposed in 1986, that capability had disappeared by 1990. I decided then that the only way the sorely needed fundamental understanding of heterogeneous catalysis might conceivably be achieved within the laboratories would be if I worked on the relevant electronic structure problems myself. I asked for resources, got them, initiated a program, and just as we had reached the point where we were able to help the experimentalists, the support was terminated, the group disbanded, and I retired. Nevertheless, along the way, I discovered in the mid 90's the beautiful way R.G. Parr and his collaborators had used density-functional theory to provide a general basis for chemical reactivity theory (CRT), and I wrote several papers extending the theory and cleaning it up.

My emphasis was on those aspects of CRT relevant to the surface chemistry of metals, extended systems with continuous energy spectra. The primary emphasis of CRT during its entire history both pre and post DFT has, however, been on atoms, molecules, and other finite systems with discrete energy spectra. A central procedure within CRT was taking derivatives with respect to electron number N, leading directly to the electronegativity, the chemical hardness, and the Fukui function, and relating indirectly to other centrally important indices of reactivity. Taking derivatives with respect to N implies that N is treated as a continuous variable, requiring the definition of density functionals for non-integer electron number. The only such definition available in the literature was based on ensemble density-functional theory. Unfortunately, EDFT leads formally to such absurdities as vanishing or infinite chemical hardnesses and to such inconsistencies as reactivities towards electron acceptors for a species with $N + 1$ electrons being the same as those towards electron donors for the same species with N electrons. All this creates a paradox because, when explicit numerical computations of the indices have been made using the standard approximate density functionals, the results made good sense. At that point, I did something Walter would never have done. I gave up.

Five or six years later, a paper by Parr and Nalewajski in PNAS on the old "atoms in molecules" problem re-ignited my interest. I had realized earlier that EDFT was irrelevant to CRT for finite localized systems. The ensemble used for a system with a nonintegral number of electrons described a physical admixture of species, each with a stable, well-defined number of electrons, and did not capture the dynamics of the valence fluctuations of a system in strong chemical interaction with its environment. My new insight was to realize that the proper density functional to use was in fact continuous in the electron number, and over time, I found powerful arguments to support that view. The result was a new formulation of the "atoms in molecules" theory and with it the elimination of all of the paradoxes and awkwardnesses of CRT, such as an undefined external potential. I am now preparing the work for publication, trying to find clear ways to put the arguments while skirting those of the mathematical morasses of the theory of functionals which are still unresolved.

The second example relates to Walter's interest in Wannier functions. With the steady progress of materials science over the years, it has become possible to prepare and to study in detail the properties of very complex materials. Wonderful things have emerged: high temperature superconductivity, colossal magnitoresistance, orbital and charge ordering, high dielectric constants, strongly piezoelectric materials, and more. Having available first-principle calculations of electronic structures and properties as well as structural and compositional phase diagrams of the materials would be most useful. However, many of the most interesting materials are complex oxides with many atoms in the unit cell, prepared not only as single crystals but as thin films or multilayers. Direct attack on the phase diagrams and on the temperature dependences of properties of such complex materials can still be beyond present first-principles computational capabilities. To surmount this problem, Karin Rabe proposed construction of a model Hamiltonian with which the computations were feasible and the parameters of which were to be obtained by mapping from first-principles total-energy computations to the model Hamiltonian. For the perovskite ferroelectrics and related materials of specific interest to her, she recognized that the coordinates of the model Hamiltonian had to be taken from the unstable normal modes of the cubic structure. So, by analogy with the electronic Wannier functions, she constructed the lattice Wannier functions of the unstable modes. Feeling pleased with herself, as anyone would, she checked the literature, only to discover that Walter had already introduced the concept and worked it out in one dimension. Finding that Walter was there first is a common experience. The model Hamiltonian approach has been very successful in the hands of Karin Rabe and David Vanderbilt and their collaborators for thermodynamic-equilibrium properties. My role has been to generalize the method to model Langevin dynamics for

Morrel H. Cohen

non-equilibrium problems and to use it to construct, with Jorge Iniguez and Jeff Neaton, a theory of pressure amorphization.

Finally, in the summer of 1997, there was a workshop at the Aspen Center for Physics on the theory of insulators. As time passed, I had heard no mention of Walter's paper proving that insulators were distinct from conductors in that in a certain specific sense their many-electron wave functions were localized. As this was the single most important paper on insulators since the early development of band theory, I gave a brief informal talk calling it to everyone's attention. Rafaele Resta and Richard Martin were at the workshop. One year later, Resta published a paper arguing that electrons in insulators were localized from a point of view somewhat different from Walter's. Two years later, Richard and his student Ivo Souza published a beautiful extension of Walter's work, bringing Walter's localization theory again to attention at a time when the field had matured to the point where its significance could be fully recognized, many years after its publication.

So, riding in Walter's wake has made my life in physics easier and more rewarding, as it has for many others. I am deeply grateful to him.

**About the Author**: Morrel H. Cohen was Louis Block Professor of Physics and Theoretical Biology at the University of Chicago and subsequently Senior Science Advisor at the Exxon Corporate Research Laboratories. He is currently Distinguished Scientist in the Department of Physics and Astronomy of Rutgers University where his research concerns electronic-structure theory, fundamental questions in chemistry, developmental biology, human prehistory, and econophysics. Contact: Department of Physics and Astronomy, Rutgers University, 136 Frelinghuysen Rd., Piscataway, NJ 08854-8019, U.S.A.; mcohen@physics.rutgers.edu

# Paris Acquaintances

Claude Cohen-Tannoudji

Laboratoire Kastler Brossel, Paris, France

I have known Walter Kohn for about 30 years. I met him for the first time in the house of my close friend André Blandin, who unfortunately died in 1983. I still remember a wonderful summer evening where we had a dinner together in a very nice restaurant outside Paris with a splendid view of the city.

Discussions with Walter are always very exiting, not only on Physics but on any other subject. His very broad culture, his kindness, his sense of humor are unique. And I would like to tell him, at the occasion of his eightieth birthday, how happy and how proud I am to be one of his friends.

**About the Author**: Claud Cohen-Tannoudji received 1/3 of the Nobel Prize in physics in 1997 for the development of methods to cool and trap atoms with laser light. Contact: Département de Physique – Ecole Normale Supérieure, Laboratoire Kastler Brossel, 24, rue Lhomond, F-75230 Paris Cedex 05, France; Claude.Cohen-Tannoudji@lkb.ens.fr; www.lkb.ens.fr/~cct/

# Walter Kohn and the UCSB Campus

France Anne Córdova

University of California, Riverside, U.S.A.

I first met Walter Kohn when I moved to UC Santa Barbara in the summer of 1996, to assume the position of Vice Chancellor for Research. As I am an astrophysicist, I was invited to join the faculty of UCSB's Physics Department. My first visit to my faculty office in Broida Hall was on a Saturday and I noticed that the door next to my office was open and someone was working there, in a nicely carpeted, bright and cozy room. I introduced myself to Walter Kohn, who was beaming broadly and welcomingly.

I discovered that Walter had made a significant mark not only in condensed matter physics, but on the UCSB campus. He was the founding director, in 1979, of the National Science Foundation's Institute for Theoretical Physics, sited at UCSB. The Institute brings together leading scientists from all over the world to address cross-disciplinary problems in physics and related areas. Walter and his successors have made the Institute (now called the Kavli Institute for Theoretical Physics) into a major research center. The elegant building housing the Institute is called Kohn Hall to honor Walter's leadership and contributions. I found out that, although ostensibly retired from the faculty, Walter was in his office every day that he was in town, doing research. He was never too busy to engage in a discussion, on any subject from cuisine to cosmology. To this day, Walter attends most faculty meetings, contributing reasoned arguments and a gentle laugh that brings clarity to vigorous discussions.

There is so much that endears Walter to his colleagues and students: his versatility of interests (for example, he is an articulate and thoughtful contributor to UCSB's program on science and religion), his advocacy for world peace, his calm but passionately delivered oratories (dare I write that he has a lot to say?), and his affection for students and support of science education.

Perhaps the most wonderful thing to happen during my early years at UCSB was Walter winning the Nobel Prize in 1998. We awoke to the news on the radio at 6:30 a.m. and the entire town of Santa Barbara was soon buzzing about it – even the fellow selling bagels at a Jack's had heard the news and was beaming it to all his early morning customers. Santa Barbara's own "nobility"! What was especially remarkable to us campus citizens was that Walter won the prize in chemistry – what a terrific statement about our touted environment of interdisciplinarity! This special aspect of the Prize brought all the science and engineering departments even closer together, knowing that each department shared the riches of its faculty throughout the university. This was later reinforced in 2000 when Alan Heeger, also a member of the UCSB physics faculty, won the Nobel Prize in Chemistry and Herbert Kroemer, a member of UCSB's engineering college, won the Nobel Prize in Physics. A previously unidentified "exclusion" principle was at work in which a UCSB faculty member only won the prize in another department!

With their charm, their involvement in all things to do with the university, and their love of people and culture, Walter and Mara made UCSB a special place for me and for everyone who was fortunate to know them.

Happy Birthday Walter! You are a friend and inspiration.

**About the Author**: France Córdova is Chancellor of the University of California at Riverside. Past appointments include Vice Chancellor for Research and Professor of Physics at the University of California at Santa Barbara, and Chief Scientist for the National Aeronautics and Space Administration (NASA). Córdova's scientific interests include multispectral research on compact binaries and space-borne instrumentation. She is U.S. Principal Investigator on the Optical-Ultraviolet Monitor Telescope on the European Space Agency's XMM-Newton satellite, launched in 1999. Contact: Chancellor's Office, 4148 Hinderaker Hall, UC Riverside, Riverside, CA 92521, U.S.A.; france.cordova@ucr.edu

# Principle and Practice

Bernard Delley

Paul Scherrer Institute, Villigen PSI, Switzerland

The first time I met Walter was on the occasion of a workshop organized by Wanda Andreoni and entitled "Electron States in Atoms, Molecules and Solids: the Point of View of the Physicist and of the Chemist," in Les Diablerets Switzerland in October 1985. I vividly remember the lively, spontaneous gatherings with Walter, Michael Schluter and some other untiring chaps in the lobby at late hours. The focus of these discussions was typically on the fundamentals of various issues in physics.

Since then, I have been fortunate to meet again with Walter on many occasions. Notable were the bag-lunch powwow's of the 'DFT-aficionados' in the Pauli room at the ETH, when Walter was on his yearly visit in Zurich. Whenever possible I came into town from the Paul Scherrer Institute for this rencontre. These lunch meetings were an excellent occasion for me to stay on discussing with Walter topics like linear scaling methods, anions or even hydrogen-bonds and to enjoy his keen insights and his infectious smile.

Once on a stroll in Cracow, we were talking about where density functional theory (DFT) may go. One point was that the accuracy of calculated energies with today's functional approximations should be improved even for seemingly normal molecules. In my opinion, the functionals derived from a nearly homogeneous electron gas would reflect too little what happens at the edge of a molecule. I was a bit deflated, however, when Walter reminded me of his passion for demonstrating guiding principles. I was really thrilled a few years later when he explained to me his 'Airy gas' model, which strives to catch the essentials of a surface with a single characteristic parameter. His way of thinking has strongly influenced me.

The connection between fundamental principles and DFT-applications in chemical industry may be stronger than generally thought. I am convinced that a non-expert in computations can answer practical questions about weakly correlated electron states, provided the automated procedures are firmly rooted in sound principles. In parallel with the work of others, I have tried to make the DFT accessible for chemical applications. I feel that these early codes have helped convince the chemistry community that the density functional is really useful in practice. We are all indebted to Walter for the discovery of this principle.

**About the Author**: Bernard Delley is senior member of the condensed matter theory group at Paul Scherrer Institute Switzerland. His research is on the electronic properties of molecular and bulk materials, as well as surfaces. He has a long-standing commitment to the development of computational density functional methods for simulations of challenging systems. Contact: Paul Scherrer Institute, Villigen PSI, Switzerland; Bernard.Delley@psi.ch; people.web.psi.ch/delley

# Thanksgiving at the Kohns

Jacques des Cloizeaux

Cointrie Saint Aubin des Grois, France

Here you are celebrating your 80$^{th}$ birthday and it's time for take stock of your life. I am certainly not going to retrace all the things you have done, I wouldn't be able to cover everything: many of your rich and varied activities, as well as those, which belong to the history of science. I simply would like to recall the first years when I knew you at La Jolla, more than forty years ago, where our deep friendship was established so naturally, and which no subsequent event has ever disrupted.

First I would like to thank you for the interest, which you have always shown for my family and me. One thing I remember in this respect – the way you gathered us, your family and mine, at your house the first time in La Jolla for a traditional Thanksgiving dinner with turkey and pumpkin pie, which touched us deeply.

Then another time during a holiday, I also remember that I was invited to your house with some other students to help plant a tree in your garden. You really didn't need help – the site had been chosen and the hole dug – but it symbolized your desire to have people participate in and share events that are important to you. Much later, when I told you that we were going to have a third child and I wanted to work during the vacation, you immediately suggested that I replace you, and since the company didn't accept your proposal, you got around the situation by having one of your friends recommend me. Those two years sped by and we returned to France.

Later, we met in different places – in France and in the States, particularly in Santa Barbara where you were directing the Institute of Theoretical Physics. Everywhere I saw and appreciated your sense of duty and your quest for perfection. For many years we only spoke in English. Nicole and I knew that you read in French, but that you refused to speak it, and then all of a sudden after one trip, to our great surprise, you started to speak in French – perfectly. That's why, knowing your interest in our "old" Europe and for our country in particular, I am writing this letter to you in French, which will be translated into English for the publication. Dear Walter, once again, I thank you for our friendship.

**About the Author**: Jacques des Cloizeaux is a retired researcher from the Service de Physique Théorique at Saclay, Atomic Energy Commission (CEA). His main field of interest is theoretical polymer physics. Contact: F-61340 Cointrie Saint Aubin des Grois, France

# Walter Kohn and UC Policy

Hugh DeWitt

Lawrence Livermore National Laboratory, Livermore, U.S.A.

In the summer of 1986 I had invited Walter Kohn to a meeting at UC Santa Cruz on Strongly Coupled Plasma Physics, and asked him to speak on the application of density functional theory to extremely high density plasmas as found in some stellar interiors. He came to the conference and gave a most interesting and useful talk. At this time I had the chance for the first time to discuss not only physics with him but also his deep concerns about the nuclear arms race. This was still during the Cold War and the Reagan Administration seemed to be pushing for new nuclear weapons and the Strategic Defense Initiative (SDI, also known as Star Wars) without end. As a long time staff member of the Lawrence Livermore Lab, one of the two US nuclear weapons design laboratories, I shared Walter's worries about the seeming runaway nuclear arms race. For several years already I had been a vocal internal critic of the manner that the Livermore Lab sold new weapons ideas to the US government. Also I was convinced that an end to nuclear weapon testing was feasible and desirable, and that the new Reagan Star Wars program was dangerous for the world. At the Santa Cruz meeting Walter and I discussed at length our mutual concerns and what we could do to try to slow down the arms race. We remained in close contact on these matters for the next several years.

Walter and I quickly found that we shared a related concern: namely, that the University of California was deeply involved in the arms race by virtue of its management of the Livermore and Los Alamos Laboratories for the US Dept. of Energy. This arrangement had begun in 1943 during WWII for the development of the first atomic bombs at Los Alamos. The contract for management of Los Alamos by UC was continued after WWII, and renewed every five years, and included the Livermore Lab as well when it was started in 1952. The UC management of the nation's nuclear design laboratories was justified as a matter of public service by the University. Walter felt strongly that it was a perversion of the mission of the University to be involved in secret classified weapons work. For my part as a Livermore Lab staff member of many years I felt that the University oversight of the labs really was a cover of respectability. The University in my view really had little to do with the lab programs. The very first UC faculty committee, the Zinner Committee, in 1970 had investigated the UC management of the weapons labs and concluded that UC's role was at best that of a "benevolent absentee landlord". Walter wanted UC to divest itself from the management of the labs, and let the Dept. of Energy find a more appropriate contractor. I initially wanted more UC involvement with the management of the labs in the hopes of promoting more broad scientific research and diversity to peace time energy programs. The issue was quite complex and had strong partisans on both sides. Walter emerged as a leading spokesman on the UC faculty for ending UC management of the labs. He wrote endless letters and attended dozens of meetings of UC faculty senates promoting his view. I helped out from the Livermore Lab with as much information from inside the lab as I could gather.

During the fall of 1989 I was at UC Santa Barbara running an ITP program on strongly coupled plasmas which Walter also participated in. This gave us ample time to develop arguments for changing the UC management of the labs. The matter was more and more timely because the removal of the Berlin wall in 1989 was the beginning of the end of the Cold War, and time for new thinking about UC's involvement in future US defense programs. Walter's outspoken public statements calling for UC to get out of running the US nuclear labs resonated with faculty members on most of the nine UC campuses. There were impassioned debates on the matter on some campuses, and on several of the campuses there were faculty wide votes calling on UC to not renew the management contract of the labs for DOE. A couple of the campuses voted in favor of divestment of the labs. Even so the effort failed in that in 1992 the UC Regents once again renewed the management contract, and the renewal has happened again each time it came up. So today UC is still deeply involved in the US nuclear weapons programs by virtue of its continuing nominal management of the labs. Walter's efforts were not in vain, however. There have been extensive changes in the DOE-UC contract for the labs. There is some increase in UC involvement and more oversight. Also the protection of internal critics and whistle blowers at the labs has been upheld. Both labs are far more open now than they were 20 years ago, and it seems that unnecessary secrecy has been reduced in some of the newer lab research programs.

I suspect that Walter will be interested and maybe amused by the recent scandal at Los Alamos that forced the top level UC administration to step in. Los Alamos had hired two special auditors to look for fraud and miss-appropriation of government property at Los Alamos. These two auditors in a few months did their job, but evidently stepped on toes of mid level managers. They were fired, and then went public about what they had found. This matter got to Congress and started an ongoing uproar, which forced the Los Alamos director to resign, and UC had to take over management of financial affairs at the lab. There is now even discussion in the DOE that possibly the weapons lab management does not belong in the hands of the University of

California after all. My guess is that this scandal will blow over, and the status quo will remain.

Walter has other talents besides being a great physicist and a peace leader. He is also an outstanding gourmet cook! My wife and I were guests of Walter and Mara at their home one weekend in the 90's, and had occasion to really enjoy some very delicious meals cooked and served by Walter.

**About the Author**: Hugh DeWitt has been a staff member of the Lawrence Livermore Laboratory since 1957, and though retired is still there as a Participating Guest. His research interests are quantum statistical mechanics of plasmas with applications to steller interiors, and Monte Carlo simulations of strongly coupled plasmas, fluid and solid, for the study of white dwarf and neutron stars. He has also for many years worked to obtain an end to nuclear weapon development and the more wasteful parts of the Strategic Defense Initiative. Contact: Lawrence Livermore National Laboratory, 7000 East Ave., Livermore, CA 94550-9234, U.S.A.; hedw@hdiv.llnl.gov

# Walter Kohn and the Australian Psyche

John Dobson

Griffith University, Brisbane, Australia

My encounters with Walter Kohn began in 1969 when I came from Australia to La Jolla, California to begin a physics PhD. I didn't end up being supervised by Walter and didn't have anything to do with Density Functional Theory in my doctoral research, but in due course I did take Walter's Advanced Solid State class. That was undoubtedly the best lecture series I've ever attended. For every topic, Walter used his unique insight to focus on the essential elements necessary for a true physical understanding, together with just enough detail to make things concrete from a student perspective. Part of the charm was in the homework. I had never thought I would enjoy homework in quite this way. There were no "plug-ins" – each problem required just a small extension of the basic theory that he had presented, ensuring that you had to understand the subject properly. Then you were led to apply your extended theory to a realistic physical situation. Finally you had to go to the recent literature to confirm your numbers against experiment. It cannot have been easy to find problems of this style that were completely up-to-date yet feasible for graduate students within a reasonable time scale. It was worth it, though: working those problems gave us a tremendous sense of achievement, and of course we really learnt our stuff.

It was in these classes that Walter had what was possibly his first taste of the Australian characteristic that is variously described as openness, directness, bluntness, familiarity or rudeness, depending on your perspective. I had heard him expound basic Density Functional Theory in class, but I had some very uneasy feelings about it. An American student would have been smooth enough to voice such personal doubts privately to him. An Asian student would probably not have dreamt of questioning the Guru under any circumstances. But the token Aussie just came straight out with his doubts right there in class, as soon as they occurred to him. My uneasiness related to issues that were subsequently termed $v$- and $N$-representability, and that were solved satisfactorily in due course (not by me!). Walter was not at all flapped by this questioning, and replied that there were indeed some significant issues involved, but that they were already on their way to resolution.

Walter got his own back some time later by putting me on the spot in one of his famous Brown Bag Lunches (see other articles). I was asked, at zero notice, to explain the work I had done with my "boss", Don Fredkin, predicting one-way propagation of collective conduction electron spin waves along a metal surface in the presence of a magnetic field. After hearing the wonderfully physical explanations Walter came up with in his graduate classes, I knew I couldn't just say, "the spin hydrodynamic equations predicted this result". So I guessed, aloud, that the one-way phenomenon was the smeared out hydrodynamic remnant of the one-way surface bouncing or skipping motion that individual electrons experience when a magnetic field is applied parallel to the surface. I don't know how plausible that was, but I noted that my audience included famous French theorist Philippe Nozières as well as Walter and various other UCSD luminaries. Daunting indeed.

After graduating with a PhD from UCSD I went to Cornell to work with Neil Ashcroft, and then back to Australia to a permanent academic job. Through a collaboration with Jim Rose I became more interested in electronic surface phenomena and hence inevitably in Density Functional Theory. So, with my family I visited Santa Barbara on sabbatical in 1989 and 1993, and interacted quite a bit with Walter as well as experimentalist Beth Gwinn. Walter and Mara got on famously with my wife Astrid and my kids William and Sarah. I also met Hardy Gross there, cementing a friendship with him and forming an interest in Time Dependent DFT that has stayed with me to this day. During this period I had become convinced that the RPA groundstate correlation energy contained the essential physics of the distant van der Waals (dispersion) interaction, a long- ranged correlation phenomenon that is absent in the conventional Local-Density and Generalized Gradient approximations. I wanted to find a DFT-like approximation for the extended RPA correlation energy. In common with many other writers in this book, I found that Walter had already had similar ideas to mine and was in fact ahead of me. He had written manuscripts with Werner Hanke, though he had published nothing up to that time.

In this period Walter got some more experience with Australian directness. Walter is famous for weighing his words carefully, to the point where even a relatively low-keyed conversation often contains some lengthy pauses. Not surprisingly, the "direct" Australian has problems dealing with this. For a long time I repeatedly made the mistake of thinking Walter had finished, and diving in with my next comment. This would elicit just the flicker of a frown, and I would realize my mistake, too late. After a while at Santa Barbara I did start to learn.

It wasn't always easy to get time with Walter, as he was much in demand. I remember one morning feeling particularly pleased with myself for catching him just as he walked into the Department. "Oh Walter", I said, "I'd like to discuss ... with you!" (I've forgotten which particular technical point was burning me up at the time). He looked a little bemused and said "Could you give me 5 minutes?" Once again, too late, I realized that I had jumped in too soon, and had cornered the poor man when he had just stepped off the bus and walked up several flights of stairs. He probably wanted to take off his jacket, put down his bag, maybe wash his hands, whatever. So I decided to blame it on national characteristics once more. "Sorry, Walter" I said, "it's just my Australian abruptness." Seeing that he had unwittingly contributed to my self-enlightenment, he gave me an enormous smile and a well-known quotation: "Know thyself!" he said. And I felt that I did.

After attending an excellent DFT workshop at the Institute for Theoretical Physics (now the KITP at the Kohn Hall) in 1994 I found myself so enthused that I arranged a somewhat similar affair at my home institution, Griffith University in Brisbane, Australia. It took place in 1996, and had a very pleasing international attendance list including Walter. He and Mara had a pretty good time in Australia, except for feeling a little cold in the sunny but windy Brisbane winter weather. Walter liked the scenery and wildlife, but commented especially on the openness, friendliness and helpfulness of the Australians he met. Unfortunately this let the cat out of the bag – now that Walter had seen some other Australians I couldn't really pretend that my own abruptness was entirely due to a ubiquitous national characteristic. Well, it worked for a while.

So, happy 80th, Walter! I hope to get the chance to grab your attention and interrupt your pauses for many years to come.

**About the Author**: John Dobson is a Professor of Physics at Griffith University, in Brisbane, Australia. He received his Bsc and MSc at Melbourne University, Australia, PhD at U. of California, San Diego, and did a postdoctoral stint at Cornell University. His interests are in condensed matter theory, particularly many-electron physics. Particular current foci are Time Dependent Density Functional Theory and theory of dispersion forces. Contact: Griffith University, School of Science, Kessels Rd, Queensland 4111, Australia; J.Dobson@sct.gu.edu.au

# Reunion in History

Gertrude Ehrlich

Hyattsville, U.S.A.

Dear Walter: I was asked to write an anecdotal statement, but I prefer the second person singular. Welcome to the club: having just turned eighty myself, I can report that becoming an octagenarian hasn't changed me significantly. What is so special about the decimal system, anyway? In the duodecimal system, your next important birthday would be four years away – and it would be written "70".

As I have said before: we were in history together. Let me review some of that history as I remember it. The Anschluss occurred in March '38, and they kicked us out of our respective schools in June '38. Many who were kicked out, even among those who survived, never went to school again. A few hundred of us met the criteria, and had the desire, to be admitted to the only Gymnasium in Vienna that could still accept Jews: the Chajes Gymnasium.

The most memorable day of the school year 1938/39 was November 10, 1938. It was the day following the night now known as Kristallnacht – a rather callous name, I've always thought, since the glass that was broken during that government-sponsored pogrom was much less important than the lives that were destroyed. We were restless in school that day – our Latin teacher, Guido ("Gedalje") Berger tried to make us keep our troubled minds on Vergil. He made a pretty impressive speech, reminding us that our brains were our chief assets, and that we must not lose our focus on learning, come what may. Director Nohel decided to release us early, in small groups, to avoid attracting attention. The next day, it was rumored that you and a fellow named Raubitschek had been arrested on your way home – but they let you go because you were just fifteen.

Ours was a lively class, with a predominant interest in science – but we enjoyed other activities, as well, such as reading plays aloud in German class. One day, our classmate Bibi entertained us with music: he had discovered a way to play tunes on his fountain pen. I think our teachers appreciated us, and we appreciated some of them. The class got smaller and smaller as the year progressed – the lucky ones got their visas early. Emigration was on everyone's mind.

Eventually, June '39 arrived and school ended. On the last day, we stood around, waiting to say our good-byes before dispersing. In particular, I wanted to say good-bye to Rudi, a mathematical star of the class, but he was surrounded by too many people. So I gave up and left, dejectedly starting to walk home, toward an uncertain future. Then someone called my name and caught up with me: it was Walt(h)er[1] Kohn, a friendly, young-looking kid I'd never spoken to before. We walked together and chatted, all the way to the Augartenbrücke – discussing our respective emigration plans. By the time we said good-bye, I felt much better: perhaps there would be a future, after all!

Fast-forward now to 1998: a woman on TV, having just announced that the Nobel prize for chemistry had been awarded to one Walter Kohn, from Austria, corrected herself to say: "from the United States, a native of Austria". "Chemistry" puzzled me since I knew you had become a physicist, but the Nobel website assured me that the winner was, indeed, a physicist. Your response to my congratulatory letter included a beautiful, self-deprecatory cartoon – and the suggestion that we hold a class reunion!

Several of the classmates who were known to have survived the Holocaust had died in recent years – including the Swedish mathematician Karl Greger, the New York engineer Paul (Bibi) Sondhoff, and Ludwig Levai who had returned to Vienna to run his family's paper factory. The Italian mathematician, Rodolfo (Rudi) Permutti, was alive but not well enough to come. Alive and ready to join you in a reunion were Ludwig's widow Ilse (Arnold) Levai, Herbert Neuhaus (a retired Chicago physician), and myself.

The four of us, along with Brigitte Neuhaus and Mara, met for dinner on May 18, 2000, on Capitol Hill in Washington. At the dinner, awed as we were by your exalted status, we were quickly put at ease by your relaxed manner and your earnest efforts to get to know us. We talked about our fellow students and our teachers – in particular, about our biology teacher Elise Deiner and our physics teacher and school director Emil Nohel, both of whom had been murdered in Auschwitz. The following day, at the National Gallery, to which you and Mara had walked from your hotel, you began by telling us that you had just seen grown people climbing around on Einstein's head in front of the National Academy of Sciences – somewhat to your dismay. You then had a friendly argument with the lady at the front desk of the Gallery regarding the newly acquired da Vinci. (Was it, or was it not, the only da Vinci in the U.S.?) When we arrived at the da Vinci, you told us the model had died shortly after posing for this portrait. I recall a few other instances during our gallery visit where the relationship between artist and model engaged your interest. You were particularly rough on a Dutch painter (was it Frans Hals?) who, in your opinion, demeaned his subjects through his representations of them. – Our best moments at the Gallery were the moments when you peppered us with questions, trying to get to know more about our lives, and making us feel worthy of your interest. Eventually, you treated us to a cab ride to your hotel

---

[1] When did you change the spelling?

where you and Mara showed us your pictures of the 1999 award ceremony, and told us many great stories. (I'll bet you are the only nobelist who ever delayed his award ceremony by a year so that his wife would be able to join him there!)

The class reunion was a wonderful idea of yours, and certainly an important event in the lives of those of us who were able to attend. Later, Rudi was said to have been thrilled by your visit to his home in Trieste in 2001. As you know, he died a few months later.

Twice since our class reunion, in November 2000 and in November 2002, you were kind enough to invite me to meet you for a brief chat at the Cosmos Club in Washington, where you were attending meetings of the Population Institute. On the first of these occasions, our chat was made even briefer by a hyper-conscientious bartender who wouldn't serve you while you were dressed like a physicist – sending you to your room to put on a proper shirt and tie!! After you returned, you and the bartender seemed to be on the best of terms – I marveled at that.

I know that, in addition to your continuing scientific work, you have been working hard to transform your Nobel clout into a force for good in the human world – which needs all the help it can get! I wish you a fruitful and active life for years to come – so, please be careful on those roller blades!

**About the Author**: Gertrude Ehrlich, a native of Vienna, Austria, is Professor Emerita of Mathematics at the University of Maryland, College Park, MD, USA. Her research interests were in abstract algebra, primarily in non-commutative, associative rings. She is author or co-author of three university textbooks, two of them in abstract algebra. She was instrumental in creating the University's statewide mathematics competition for high school students. For several years, she served as an associate editor of the American Mathematical Monthly. Contact: 6702 Wells Parkway, Hyattsville, MD 20782, U.S.A.; gehrlich@wam.umd.edu

# For Rappa on His 80<sup>th</sup> Birthday from Terry

Josef Eisinger

New York, U.S.A.

It so happened, that Mara called me to ask for a contribution to this opus right after two graduate students from Vienna had interviewed me as another "émigré physicist". Among many other questions, they asked: who and what had been the most important influence on my becoming a physicist?

"Now that is an easy one!" I said and told them about our meeting in a British internment camp; how you, having attended school in England, had been seriously infected with love of physics; and how I, after working as farmhand and dishwasher had been peculiarly susceptible to your ideas; and how, by the time we reached Canada, you had passed the infection on to me. I have never ceased being grateful for that!

Walter 1941

Amid the irrational and insecure world in which we found ourselves at the time, physics did indeed provide a refuge of order and reason for us, a role that science plays for me to this day. Numberless students must have employed you as their "role model" since then, but I will always claim the distinction of having been the first to do so and of having been your friend since those distant days when you sported this abundant shock of hair and your eyes twinkled – even as they do today!

And so I raise my glass of beer (though not comparable to Mara's, c.f. page IX) and proclaim:

The world is often wracked with pain,
If it isn't George Bush, it's Saddam Hussein,
If not Prof. Burton[1], it's Edward Teller
    Or some other unspeakable feller.

But let's not despair or pull out our hair,
For I know a person with an uncanny flair
For righting wrongs with cool-headed reason
    And fighting for justice whatever the season.

He travels the world preaching physics and peace,
To Cuba for chili, to Zurich for cheese,
And leaves unturned not a single stone,
    That redoubtable indefatigable Kohn.

His friends and family all stand in awe,
And while his activities don't break the law,
They are somewhat unseemly and quite contrarian
    For one who is now an octogenarian.

---

[1] Prof. Burton, for reasons of his own, had kept Walter from entering the Chemistry building at the University of Toronto, labeling him a 'security risk'.

But enough of all this useless chatter
Whether eighty or ninety, what does it matter?
Long live this physicus extraordinaire
    In health and with vigor, bis Hundert und mehr!


**About the Author**: Josef Eisinger is professor emeritus of biophysics at the Mount Sinai School of Medicine, New York. His eclectic research career began at MIT where he used atomic beam methods to study nuclear structure. After joining Bell Labs he developed magnetic resonance, emission spectroscopic, and energy transfer methods for investigating the structures of nucleic acids and proteins and co-founded the Molecular Biophysics Department. He used fluorescence microscopy to investigate biomembrane fluidity and many other biological systems, some leading to useful medical applications. His non-physics writings range from history of medicine to musical biography. Contact: 197 West Houston Street, New York, NY 10014, U.S.A.; jeisinger@att.net

# Meeting Walter Kohn in Internment Camp

Ernest L. Eliel

University of North Carolina at Chapel Hill, Chapel Hill, U.S.A.

As a consequence of the ill-advised internment of Austrian and German refugees in Great Britain in May/June 1940 (ordered by Winston Churchill: "Collar the Lot"), Walter Kohn and I met in internment camps in Britain and, later, Canada where we were shipped in July, 1940. We often talked, and I was sometimes piqued by the realization that though Walter was the younger of us (by over a year), he was clearly the smarter. And although the spirit among our group of young internees was generally upbeat, we sometimes had fights – verbal and occasionally physical. (I still remember with horror throwing a bowl of soup in Walter's face after an especially provocative argument – a nasty action for which I now – 62 years later – sincerely apologize.)

I believe our disagreements resulted from the fact that Walter was a conceptualist and theoretician and I was a pragmatist and experimentalist. (I had resolved to become a chemist at the age of 12.) Occasionally the argument degenerated into the foolish question as to whether physics or chemistry was the superior science. (Of course, one can make arguments for either!) In a way, Walter had a situational advantage: it was difficult to carry out experiments in an internment camp but it was quite possible to read books and think. (I tried to catch up with him in gaining an understanding of early 20th century physics, but was not as good at it as he was.) As a matter of fact I believe that, while many of the important scientific discoveries of the second half of the century (such as NMR and lasers) were made by physicists, some of the most useful applications thereof turned out to be in chemistry (and biology). Thus, while at the age of 18 Walter may have insisted that physics was the superior science, it seems entirely appropriate that he eventually won the Nobel prize in chemistry for his pioneering development of density functional theory. Thus, after 62 years, it appears that the silly old fight has ended in a draw!

In congratulating Walter on his special birthday and wishing him long continuation of his active professional life in health and happiness, I remember our joint stay in Camp B, Ripples, N.B., Canada many, many years ago with fond if bittersweet feelings.

**About the Author**: Ernest L. Eliel is W. R. Kenan Jr. professor emeritus in the Department of Chemistry of the University of North Carolina at Chapel Hill. His research has been in the area of stereochemistry and conformational analysis as well as stereoselective organic synthesis. He is author of "Stereochemistry of Carbon Compounds" (McGraw-Hill, 1962), a widely used text, coauthor of "Conformational Analysis" (Wiley, 1965), "Stereochemistry of Organic Compounds" (Wiley, 1994) and "Basic Organic Stereochemistry" (Wiley, 2001) and co-editor of "Topics in Stereochemistry" (1967–94). Contact: Chem. Dept., UNC, Chapel Hill, NC 27599-3290, U.S.A.; eliel@email.unc.edu

# Meeting Walter Kohn

Gerhard Ertl

Fritz-Haber-Institut der Max-Planck-Gesellschaft, Berlin, Germany

When I visited Santa Barbara in December 1981, I was also invited to Walter Kohn's tea table. His group of students and postdocs was assembled around a large table, and everybody was holding a typewritten sheet of paper in his hands. Its purpose became soon evident when Walter Kohn started to sing with loud voice: "Oh du fröhliche . . . ", and all his coworkers tried their best to join him. German Christmas carols were the last thing I would have expected at this occasion!

Recently I met Hans-Joachim Schellnhuber, who is now director of the Potsdam Institute for Climate Impact Research, and it turned out that he was working in Santa Barbara around this time. When I mentioned to him the story about the Christmas carols, he exclaimed: "O, I remember this situation well. This was Walter Kohn, with whom I worked then as a postdoc!" Small world!

**About the Author**: Gerhard Ertl is director of the Department of Physical Chemistry at the Fritz-Haber-Institut der Max-Planck-Gesellschaft and adjunct professor at the Free University, Humboldt University and Technical University Berlin. His research is concerned with the structure and reactivity of solid surfaces, including spatio-temporal self-organization, ultrafast dynamics, heterogeneous catalysis and electrochemistry. Contact: Fritz-Haber-Institut, Faradayweg 4-6, D-14195 Berlin-Dahlem, Germany; ertl@fhi-berlin.mpg.de; www.fhi-berlin.mpg.de/pc/pc.html

# Impressions of Walter Kohn

Helmut Eschrig

Leibniz Institute for Solid State and Materials Research, Dresden, Germany

## Dr hc mult, mult =: mult+1

On October 21, 2002 Walter was given a honorary doctorate by the University
of Technology in Dresden, Germany. Naturally, all of our university felt honored by Walter accepting the degree. (Among the eleven honorary doctorates

During the awarding ceremony for the honorary doctorate on October 21, 2002 at
the TU Dresden. *Left:* Magnifizenz Prof. Dr. A. Mehlhorn, Rektor of the university;
*Right:* Walter (photo: TU Dresden)

Presentation of the document (photo: TU Dresden)

awarded by the TU Dresden in the past four years the two most outstanding are Walter and Kofi Annan.)

We had four really exciting days with Walter at our faculty, the Max-Planck Institutes for Chemical Physics of Solids and for Physics of Complex Systems (the latter was designed a few years ago with the ITP Santa Barbara in mind) and at the Leibniz-Institute for Solid State and Materials Research. Not only were we deeply impressed by the wise speech of Walter at the ceremony, urging us to have the future in mind with caution, he also amazed us with his never getting tired when the day became long with intensive discussions on science as well as on people.

## Walter's Accuracy of Thinking

One benefit of contact with Walter is a continuous learning from his high standards, to improve the precision of one's own statements. Two examples:

Occasionally I had to go through a text of mine with Walter, which contained one of the commonly used phrases: "In this situation, everybody recommended ...". With his so familiar smile Walter asked: "Are you sure that really everybody recommended that? Shouldn't you better say that many recommended it." Of course he was right.

In another passage of the same text I intended to report an event from history, I did it in affirmative phrases. Same smile of Walter. He asks me: "Are you sure that it is true?" Me: "I read it in print." He: "But you know that not everything in print is true." Again I had to confess that he was right and I finally started the text with: "It is reported that ..."

## Walter's Humor

Everybody knows Walter's earnest relation to science. Nevertheless, he can also look at it with his eyes of smile. He was the founding director of the Institute for Theoretical Physics at Santa Barbara (now Kavli Institute for Theoretical Physics) and was its first director. With his relation to the Niels Bohr Institute in Copenhagen, going back to his early postdoc years shortly after World War II, he considered carefully the structure of that institute when participating in designing the ITP Santa Barbara. Nevertheless he likes to tell a story which goes like this:

The Niels Bohr Institute happened to acquire a project from the Rockefeller foundation. When a high-ranking officer of the Rockefeller foundation by chance came to Copenhagen, he visited the Niels Bohr Institute and asked Bohr to show him the institute's most advanced device. Bohr, after hesitation, said: "We don't have devices". The officer, very astonished: "But then, what are you doing." Bohr, after minutes of thinking: "We try to explain to each other what we do not understand."

**About the Author**: Helmut Eschrig is scientific director of the Leibniz-Institute for Solid State and Materials Research, Dresden and professor for Solid State Physics at the University of Technology Dresden, Germany. He is a scientific member of the Max-Planck Society and a Member of Leopoldina and of the Saxonian Academy of Sciences of Leipzig. His research concerns density-functional theory, theory of magnetism and theory of superconductivity. Contact: IFW Dresden, P.O.B. 27 00 16, D-01171 Dresden, Germany; h.eschrig@ifw-dresden.de; www.ifw-dresden.de/~helmut

# Reminiscences of Walter

Peter J. Feibelman

Sandia National Laboratories, Albuquerque, U.S.A.

I was not quite 21 when I arrived at UCSD for graduate school, determined to sign on with the most eminent physicist in the department. That turned out to be a real problem – seemingly every professor there was a superstar. The scuttlebutt among the students (there were just 300 of us on the 3-year-old campus, all male, I might add) was that of all the theoretical luminaries, Walter worked his advisees the longest and hardest, which scared off all but the bravest. I had seen Turandot. I was no Calaf.

Though not Walter's student, I benefited greatly from his presence on the La Jolla campus. I audited his Solid State Physics course, perhaps intuiting that that area was my ultimate destination. I got to know his marvelous post-docs, Maurice Rice and Lu Sham. I attended the seminar at which he reported the work he'd done in Paris with Pierre Hohenberg.

Did I have any idea of the impact this work would have on Condensed Matter Physics? No. On my career? Again, no. In retrospect I wonder if Walter himself foresaw that the H–K theorem would be the basis for thousands of journal articles and enormous strides toward understanding the structure of matter. I guess he did. But I, certainly, was too ignorant to realize I should drop everything and apply to work with him. I was at the "right place at the right time" – every student's dream – but too unformed to take advantage.

One of my greatest pleasures at UCSD, far from family (in New York City) and traditions, was Seder at the Kohn's. The warm atmosphere, the cultured environment did a lot to counter the sense that I was an exile – perhaps just an expatriate – in paradise (La Jolla in the 1960's was paradise!). Oddly, my strongest Seder memory is of Kazumi Maki and me singing Schubert lieder, with Walter's wife at the piano. What fun! What a challenge trying to decode Maki's mysterious German.

I don't recall giving a seminar at UCSD, but attended many. Whenever Walter was present at one, glances would skitter in his direction searching for signs that "the smile" was about to appear. Walter's eyes disarmingly atwinkle and his grin as broad as could be were sure indications that the speaker was in deep trouble. Probing questions were certain to follow, and after the talk, shaking heads. Ay, caramba!

Much later in life, two decades later, I spent five weeks at the ITP in Santa Barbara. One day, Walter asked me if I had invented the surface optical response "$d$-functions" all by myself (I had), and told me how much he liked my review article on Surface Electromagnetic Fields. What a thrill for me! Noticed and complimented by a culture-hero, and no smile.

**About the Author**: Peter J. Feibelman is a theorist in the Surface and Interface Sciences Department at Sandia National Laboratories. For his early work on surface electromagnetic response he received the American Physical Society's 1989 Davisson-Germer Prize. Feibelman is co-discoverer (with M.L. Knotek) of a basic mode of stimulated desorption. Since 1978, he has been using density functional theory to gain insight into surface structure and dynamics. Feibelman wrote the popular book, "A Ph. D. is Not Enough: a guide to survival in science." In 1996, he received the American Vacuum Society's Medard W. Welch Award for outstanding research in Surface Science. Contact: Sandia National Laboratories, Albuquerque, NM 87185-1413, U.S.A.; pjfeibe@sandia.gov; www.sandia.gov/surface_science/pjf

# To Know How to Count ...

Michael E. Flatté

University of Iowa, Iowa City, U.S.A.

Near the end of my very first research project, with the new contribution finished and an example calculation desired, I needed only to determine a minor characteristic of the vibrational modes of a slab – the density of states of modes of mixed longitudinal and transverse character through reflection from a free surface. My intuition failed me at this point, and I went to seek guidance from Walter. He listened patiently to my concerns, and looked politely at a couple of pages of notes I had prepared and a few pages from a book I showed him. My point was that there were conflicting results in the literature, and I did not know which one was right. Walter thought briefly. Then he said, with deliberate care – both respectful and firm – "I am not going to help you with this. I think you can solve this yourself." He then continued by telling me of a statement that an eminent physicist of an earlier generation had made (I forget who – it does not matter to me, for Walter is my source): "The single most important ability for a theoretical physicist ... is *to know how to count*." In typical style he then restated that the important aspect was not doing the counting itself, but having the *ability* to count.

This was not a challenge I could decline. I did not return until I had sorted out the problem completely. Later I observed that this theme had a special place in Walter's heart – I remember he remarked with pride that the great achievement of density functional theory was the reduction of a problem with $3N$ degrees of freedom (the coordinates of the $N$ particles) into a problem with only 3 degrees of freedom (the coordinates of the density). The comment of a "counter" indeed! From this exchange I learned about motivation, respect, kindness, and precision.

Moments like these, when I was gently pushed towards the right path, convinced me that Walter was the best adviser possible for me. Now when I am faced with a challenging supervisory situation I often take a few seconds (or minutes ...) and think of what Walter would do. And once or twice the circumstances have been just right, and I have told this story.

**About the Author**: Michael E. Flatté is an Associate Professor of Physics at The University of Iowa. He received his Ph.D. working with Walter Kohn in 1992 at UCSB. He has worked on thermal energy atom surface scattering, the vibrational properties of superconductors, the local influence of impurities on superconducting coherence, the design of semiconductor superlattices for lasers and detectors, and the nature of spin coherence in semiconductor nanostructures. He spent the 2000-2001 year on sabbatical back at UCSB meeting old friends and making new ones. Contact: Dept. Physics and Astronomy, U. Iowa, Iowa City, IA 52242, U.S.A.; Michael_flatte@mailaps.org; www.ostc.uiowa.edu/~frg

# My Friend and Colleague Walter Kohn

Jacques Friedel

Académie des Sciences, Paris, France

I first met Walter in the summer 1953 at a Gordon conference where, after my first trip across the Atlantic, I met the cream of the people then active in the electronic structure of solids. I had used, three years earlier, his technique with Rostoker to analyze the conduction band of polyvalent metals such as Mg or Al, to understand how and how much they deviate from the free electrons model in their cohesive properties. By 1953, my interest had switched to impurities in metals; and it is only with André Blandin that, after Walter himself and John Ziman, I would come back to this question, using a simplified perturbation technique.

Walter was then for me a young but established theoretician, with broad interests from nuclei to solids. We met a number of times in Paris, Brussels and especially at the 1956 Varenna Conference, where a number of us brought our young families and formed lasting friendships. This was helped by the fact that, staying through all this summer school, we had plenty of time to see each other's and were not encumbered with too many students!

It was then natural for my first PhD student, Emile Daniel, to follow Walter as a postgraduate to La Jolla, where he was arrested by the police for walking at night along the road to San Diego! André Blandin, my second PhD student, followed Pierre Gilles de Gennes traces to Charlie Kittel in Berkeley; but as he had lost his two initial races with Walter's own PhD students (on the Landau diamagnetism of impurities and on NMR in Cu base alloys), this Californian stay was the time when Walter and André became firm friends. This started a regular fashion for young theoreticians of Orsay to come as postgraduates to Walter or Charlie.

Walter started visiting Orsay on a regular basis soon after we moved there – one of the first times with Hohenberg, just after his work that would lead him to his Nobel Prize. We found renewed common interests in the field of surfaces, where Walter Kohn and André Blandin attended a CNRS meeting in Lille, and later when Denis Jérome, after staying with Walter, started his search for low temperature excitonic insulator phases. Walter's early visits were found most stimulating by the lab in Orsay. This led André Blandin, then a young professor, to propose Walter for one of the first Doctorates Honoris Causa in the newly created University Paris Sud (Orsay), a move that pleased the many friends and visitors Walter had at the Atomic Energy in Saclay. In more recent years, it is Chuck Sommers who, with the help of his computer, has kept most regularly the contact with Walter; and I am thankful for the opportunity he thus offers me to state how much I have derived from knowing Walter.

Walter had an uncle who lived in Lyon since before the war. And we often had his visits, on his way to or from Lyon, in our Yonne country place at Palteau. Our children were often there at the time, and sometimes he came with André Blandin. It was during one of these visits, on return from a drive through the autumn colored vineyards of the Auxerrois, that Walter invited me to a stay in Santa Barbara, where he was in charge of the newly created Institute of Theoretical Physics. Walter and Mara were in fact wonderfully generous of their time to make this stay a complete success, including the lending of their car to allow me to explore the coastal road up to Berkeley and back. The working of the Institute was very striking by its informality as well as by the high calibre of its participants, in a range of subjects that only somebody like Walter could have initiated. This stay in Santa Barbara helped to strengthen ties which, in a way, could be thought of being prepared when, in 1940, Mary was helping German speaking and mostly Jewish refugees, pushed out of the South Coast of England, to settle uneasily in a workhouse, north of Bristol, in the uncertain hope that the younger members of their community would, like Walter, be allowed to leave the Isle of Man.

But Walter is not somebody to reminisce too much on the difficulties of his past, when in Paris, his preoccupations in recent years have been much more on how to save UNESCO, on the best way – Spanish or English? – To teach young Mexican immigrants in California, or on how to keep a purposeful contact with young American students.

With his Nobel prize, Walter has no doubt now more opportunities to air his views, but perhaps less time to think quietly at home!

**About the Author**: Jacques Friedel is Professor Emeritus at l'Université Paris-Sud Orsay. Born 11th February 1921. Elected member of the French Academy of Sciences the 17 of January 1977. In the last few years also interested in high temperature superconductors and in one-dimensional low temperature organic superconductors. Contact: Académie des Sciences, 23, quai de Conti, 75006 Paris, France.

# Working with Walter

Michael Geller

University of Georgia, Athens, U.S.A.

I am delighted to be able to wish Walter a happy 80$^{th}$ birthday. Here I will briefly explain what it was like to be his graduate student.

## The Morning Paper

I came to Santa Barbara in 1989 specifically to work with Walter. I had visited him a few times before then, and we always had wonderful discussions in our meetings. I would ask him what problems he was currently interested in, and he would explain them to me in great detail and with his characteristic clarity.

During my first year in Santa Barbara I lived downtown with my girlfriend Robin (who is now my wife) and took the bus to campus. In those days I held the mistaken view that every physicist should read Physical Review Letters cover-to-cover, so I would try to read PRL on the bus. One day Walter sat down next to me and said "I see you're reading the morning paper." That morning I was reading an article about disordered systems, so we talked about that. He confided that some loose ends in the theory of Anderson localization have always bothered him. It turned out that we often took the same bus, so we had many similar discussions.

Walter doesn't read much physics literature himself anymore. Countless times I showed him journal articles that I thought might be relevant to our research. Hesitantly, he would look at it, but I could tell that it was more out of setting a good example for me than thinking he would learn something. After a paragraph or two his demeanor would change. "It's not clear, it makes unnecessary assumptions, and there is nothing new in it," was often his response.

## "Take Five Minutes"

During my first year of graduate school I took the condensed matter physics course taught by Alan Heeger the first quarter and Walter the second. Walter had us fill out questionaires detailing our prior condensed matter/solid state coursework. He wanted to be able to assume that the elementary aspects of the subject were understood so that he could cover more advanced material. A few of us, including me, had not taken an undergraduate condensed matter course and so were required, as part of Walter's course, to take an oral exam during the quarter on the 20 "essential" chapters of Ashcroft and Mermin's text. I studied very hard for this exam and worked through all of the homework problems from those chapters. He asked me a few questions about Brillouin zones and magnetism, and I passed. A classmate failed because he did not know the band structure of aluminum. According to Walter: You do not understand metals if you do not understand aluminum.

In addition, we wrote extensive term papers on some aspect of condensed matter theory. My paper was on spin glass theory, and I included Parisi's solution of the Sherrington–Kirkpatrick model, and discussions of replica-symmetry-breaking, ultrametricity, and so on. Walter returned the term papers to us during the next quarter, long after the course was over and grades finalized. However, he still required many students to make specific improvements to their papers and return revised versions to him!

I began working with Walter the summer following my first year. The next year he taught graduate condensed matter again, but covered some different material, so I sat in for those lectures. One day while discussing spin glasses he tells the class that "we are fortunate to have an expert on spin glasses with us today." We all turned around looking for that expert. Walter turned to me and asked whether he covered all of the essential ideas, and, regrettably, I pointed out that he did not fully discuss the concept of frustration. He handed me the chalk and asked me to "take five minutes." I did my best, but I was so nervous that afterwards I could not remember anything I had said.

## "You've told me nothing"

Walter was an extremely demanding advisor. People are often surprised to hear this, because he has such a kind and gentle personality. But when the office door closed for a meeting, he became a tiger. There was no opportunity to say "well, I got stuck" and he was rarely interested in the details of a calculation. He wanted the answer right up front, so that we could spend our time together planning the next step.

I can recall with fondness the many times I was really excited at having solved some aspect of the problem we were working on, and I marched into Walter's office, full of energy, and gave what I thought was a clear and concise chalkboard presentation about it. He sits there listening quietly, and I am thinking that he will ask me to write it up for publication. But before the chalk dust had time to settle, he said "You've told me nothing"! I kid you not. His standards are extremely high, and it is rare that he thinks a new theoretical result – coming from his students or anyone else – is highly significant.

Michael Geller

# Publishing, or Not, with Walter

Finally, I come to perhaps the favorite subject of Walter's former students and postdocs – finishing a project with him and publishing. This does not happen very often, because as I mentioned, his standards are extremely high. My thesis on compositionally graded semiconductors eventually led to three publications,[1] so in that respect I guess I did pretty well. He spent countless hours with me patiently teaching me how to write.

Once I asked Walter why it was so important to clearly explain the logic behind a calculation, as the final result is usually insensitive to this. He responded that "Clarity in writing is everything: As theoretical physicists our job is to explain why or how some phenomena happens, and the better we do this the better the work is." Readers of Walter's scientific work know that he follows his own advice.

Walter, I have so many wonderful memories from those days. I will always cherish the time we spent together. Thank you and Happy Birthday!

**About the Author**: Michael Geller is an Associate Professor in the Department of Physics and Astronomy at the University of Georgia. He is a condensed matter theorist interested in strongly correlated electron systems, mesoscopic and nanoscale mechanical systems, and vortex dynamics in superfluids and superconductors. Contact: Department of Physics and Astronomy, University of Georgia, Athens, GA 30602-2451, U.S.A.; mgeller@physast.uga.edu; www.physast.uga.edu/~mgeller/group.htm

---

[1] M.R. Geller and W. Kohn, Phys. Rev. Lett. **70**, 3103 (1993). M.R. Geller and W. Kohn, Phys. Rev. B **48**, 14085 (1993). M.R. Geller, Phys. Rev. Lett. **78**, 110 (1997).

# A Chinese Portrait of Walter Kohn

Antoine Georges

Ecole Normale Supérieure, Paris, France

When I think of Walter Kohn, I see a bridge. The pillars are made of solid stone, and have their deep foundations in the most fertile grounds, which shaped so many parts of our contemporary landscape: the world of the middle European Jewish intellectuals. The upper part of the bridge is all high-tech materials, and reaches to the blue sky of California, this uniquely creative place that has shaped our present and future for almost half a century. It is an awesome bridge, which connects two worlds, and it is very alive – being a very *human* bridge – it continues to grow and change shape and give access to new territories.

**About the Author**: Antoine Georges is conducting research in condensed-matter theory at Ecole Normale Superieure (Paris), and teaches at Ecole Polytechnique. His scientific interests are in strongly correlated electron materials, electronic structure, and the effects of interactions in mesoscopic devices. He recently co-organized a workshop at KITP-Santa Barbara. Contact: LPT-ENS, 24 rue Lhomond, F-75231 Paris Cedex 05, France; georges@lpt.ens.fr

# Memories of Walter Kohn in Santa Barbara

Hardy Groß

Freie Universität Berlin, Berlin, Germany

Walter and I first met at a density functional meeting in Alcabideche, Portugal. At this conference, I presented some results on the foundation of time-dependent density functional theory which I had just completed together with a young student, Erich Runge. Although, at first, not entirely convinced that our results were correct, he got sufficiently curious to invite me to work with him in Santa Barbara. Our work developed so wonderfully that, after a NATO fellowship and a regular postdoc position, I decided to spend a subsequent Heisenberg fellowship in Santa Barbara as well. Altogether, I had the privilege of being and working with Walter in Santa Barbara for 6 years. Countless happy memories of this time come to my mind. His passion for science, his tireless fight for human values, his marvellous ability to enjoy life, all this infected me for ever.

## The Veranda

The parties on the Kohns' veranda are unforgettable. The marvelous view on downtown Santa Barbara and on the harbor, the beautiful garden (notably the loquat trees with their delicious fruits), Walter's mastership in telling stories and anecdotes and, last but not least, his talent of cooking made each of these parties a cheerful and memorable event. To my great embarrassment, I once almost ruined the event. It was the first invitation to the Kohn's home, just after my arrival to Santa Barbara in October 1984. I felt I should bring some flowers. But they should be something really special. So I got the splendid idea to take a short hike to the Santa Barbara foothills (which, until today, is one of my favorite places on earth) to pick a bouquet of wild flowers myself. It was a beautiful bouquet, some plants with the red and yellow colors of the beginning fall. What I did not know was that this beautiful red-and-yellow colored plant was called poison oak (a rather dangerous plant for many people, as I learned later). I will never forget this split-second of desperation in Walter's eyes. The bouquet disappeared in an undisclosed location in the house and then Walter gave a cheerful discourse on the beauties and dangers of the Santa Barbara foothills. His sense of humor is truly unbeatable.

## The House

Another aspect makes the Kohns' home a rather unique place: The overwhelming treasures of artwork. Many of the pieces come from Mara's father, Roman Vishniac, a famous photographer who had been a passionate arts collector. Walter loves arts; but he is not just a lover of arts, this love comes deeply from his heart. One of the many extraordinary pieces sticks to my mind, a small tapestry, similar in style to the ones exhibited in the Musée Cluny in Paris. This particular tapestry, a piece of very special beauty, is the one Walter loves most. He once told me that in 1990 when he was evacuated from his home because of the big Santa Barbara fire, the only thing he took with him was this tapestry.

## Working with Walter

Walter tremendously enjoys the interaction with junior scientists. Working with him is not only enormously inspiring but also great fun. What I admire most is his deductive way of thinking: While each step of his reasoning is almost offensively simple, miraculously, after 10 or 20 steps, one is guided to a totally unexpected result, unexpected of course only for the audience, not from him. His fabulous intuition would usually tell him the final result from the start, but he would always insist on a proper derivation. By a discussion with Walter, a project would usually make a big leap forward. However, this did not mean that the project would get finished more quickly. On the contrary: Normally, his line of reasoning would have interesting side tracks which, he insisted, had to be understood in complete detail. Remarks like "This looks like an interesting point in its own right, why don't you work it out over the weekend?" were the most feared ones among his young co-workers: By experience one would know that a side track of this kind, typically, would correspond to the work of a whole Ph.D. thesis (or two). At first that was great. But then the next discussion on one of these side tracks would have another side track and, over time, the number of fascinating projects would grow to a daunting number. And his phenomenal memory would never forget a single one ....

Walter, the 6 years in Santa Barbara were the most fascinating time of my life. Thanks for being my friend and my mentor. I owe you more than words can say. Happy Birthday!

Hardy Groß

**About the Author**: E.K.U. (Hardy) Groß is a professor of theoretical physics at the Freie Universität Berlin. He has done research in condensed matter theory, especially density functional theory, relativistic effects in solids and the interaction of strong lasers with matter. Currently he is most interested in optimal-control theory of femto-second phenomena and in developing first-principles methods describing the superconducting state. Contact: Institut für Theoretische Physik, Freie Universität Berlin, Arnimallee 14, D-14195 Berlin, Germany; hardy@physik.fu-berlin.de; www.physik.fu-berlin.de/~ag-gross

# Thinking Back

Werner Hanke

Universität Würzburg, Würzburg, Germany

When I reflect on the many encounters I have had with Walter over more than thirty years now, numerous situations reappear, which at first glance are quite different, ranging from scientific discussions, social events to just funny encounters. But there is clearly a common denominator, i.e., that thinking back and recollecting is a source of pure joy. This joy of "thinking back" must be also one of the reasons for the success of this Walter Kohn Book and certainly is the reason for my extensive story.

But let me start from the beginning, which was in October 1972, when I arrived at UCSB in La Jolla as a young and (over)-excited postdoc. In a way, I had "met" Walter scientifically many times before in the then popular books such as Kittel's, Ziman's or the Elementary-Excitation book by Pines: Very often the "Kohn" would pop up, for example, in connection with the Kohn effect, which was very popular at that time. I had also started reading a series of papers by Walter and J.M. Luttinger on "real-world" many-body physics, i.e., their effective-mass-description in semiconductors.

On October 5, during my first Californian lunch I was introduced to Walter by Lu Sham, who was then my postdoc adviser. We marched together with other postdocs to the famous "coffee hut". On Walter's 75th birthday I have already told the story of what happened there, so I will cut a long story short: I was so excited and nervous sitting next to Walter that I spilled my pea soup onto his black trousers. Walter was just very relaxed, understanding and, very friendly, refused my offer to take it to the cleaners. Back to the department, I spotted Walter going to the men's room rather often. I got suspicious and, sure enough, Walter was using the famous yellow paper not for drying his hands but for rubbing the stain. With each trip it got larger and larger and more yellowish. I didn't dare to interrupt him on his more and more frequent trips until the evening provided a natural cut-off. So silly as the incident may sound, it is in line with Walter's aim for perfection, for flawless presentation and for the painstakingly precise way he addresses almost every issue. This was my first day with Walter, which, quite unexpectedly, was the beginning of a long friendship.

I remember La Jolla as a fantastic place for any scientist, but especially so for a young postdoc. It had attracted – mostly guided by Walter – some of the most original and exciting minds, such as Bernd Matthias, Harry Suhl and George Feher. All the German postdocs, who had the privilege to come to this most exciting place, have unforgettable memories of that time: we experienced an open welcome, help and even friendship in a place where some of the best physicists worked, all of whom had been expelled and deported in one bitter way or another from their German and European homes.

One celebrated La Jolla institution, we all profited from, was the famous "Friday lunch", a scientific gathering in a very informal way. It was a display of Walter's authority and well-functioning leadership: everyone not just had to – but also wanted to – come, even colleagues who did not seem to fit in perfectly well, like B. Matthias. In his "bouts" with Bernd, Walter was at his very best: When Bernd would try to get Walter on thin ice with sometimes provocative ("the periodic table just explains ...") oversimplifications during a dispute, Walter would react with an amazing display of patience explaining his point of view.

During the first part of my stay, I worked with Lu Sham on linear-response theory and elementary excitations in such solids, where the then much-used homogeneous electron gas description did not apply. We were especially interested in local-field and excitonic effects in transition metals, semiconductors and insulators. More or less from the beginning, Walter showed substantial interest in this work. This was partly due to his own interest in the localizability properties of Wannier functions (which we used as a possible basis set) and partly probably due to his earlier work with Luttinger on the effective-mass theory in semiconductors. In the last year of my stay in La Jolla, Lu went on a sabbatical and I had the privilege of directly working with Walter and his students. I was deeply impressed by the careful and precise way he explained the most complex issues to me and the other students. Yes, you had to – and until today you have to – be patient: the sentences are usually not short; it is like constructing a building: there is a basis, then a scaffolding of the middle part and, finally, you reach the top. It takes time, but it is a logical construction, simple and clear.

Towards the end of my stay in La Jolla, Walter was spending a sabbatical in Seattle. He invited me and I could stay in his house during my visit. There, I learned more about him, about his hospitality and about his many interests. I don't want to dwell upon any of the innumerable cultural interests, which, I am sure, are dealt with in many articles of this book. Just a completely unexpected example from more recent times: During a visit to Santa Barbara, I told Walter that I got very interested in the game of golf. I would have expected anything but not Walter telling me the correct average speed with which Tiger Woods hits the golf ball with his driver.

Later in the seventies, when I was back at the Max-Planck-Institute in Stuttgart, we had a particularly nice summer event every year, i.e., the "Parisian Summer", which was provided by the CECAM Institute in Orsay.

Paris with its innumerable possibilities provided an unforgettable background and "Bohemian life style". Clearly, this meant sampling cultural life, numerous exhibitions, music events etc. It also meant meeting Walter's close friends Ch. Sommers, A. Blandin and many others. On the weekends, we sometimes made trips to the country. I vividly remember a trip back to Paris from Normandy:

I was driving the car, when a thunderstorm got up and soon it was pouring with rain. I realized that Walter was sitting silently and seemingly a little afraid beside me. To soothe his worries, I said something like "Walter, just relax, I have some practice on bad roads. As a young man, I loved car racing and participated in rallies." I remember Walter saying with his typical dry sense of humour: "Good for you, so you think I should relax?"

Scientifically, I learned in these summers a lot about solid-state theory in general and, in particular, about non-local density-functional approximations. For example, we studied and worked out how to get the correct image-potential, i.e., long-distance behavior and Van-der-Waals forces, from a non-local theory.

During the Parisian times, I succeeded in inviting Walter (and later Mara and Walter) back to Germany for the first time, which was an unforgettable event for the Max Planck community in Stuttgart and for me.

My marriage in Santa Barbara in 1994

In the summer of 1980, I had the opportunity to work again with Walter in Santa Barbara for a longer period of time. Walter had appointed L. Falicov, B. Maple and me to run the first longer, i.e., half-year workshop at the ITP. It was devoted to "mixed-valence" compounds. At that time, these compounds posed a rather new challenge to many-body theory, i.e., one has to deal simultaneously with strongly correlated, localized electrons and rather "conventional", delocalized electrons. The setting of the ITP and Walter's deep interest in these materials provided a unique opportunity to make progress. I have profited from this experience up to today in our thrust for a better microscopic understanding of the high-$T_c$ superconductors. Many nice memories spring up, especially the warm hospitality provided by Walter and Mara in their new home in Santa Barbara.

I have enjoyed the opportunity to come back to Santa Barbara many times, usually in the summer holidays. This has developed into a more or less regular habit, mostly due to a close collaboration with D. J. Scalapino on the high-$T_c$ theory. In all these years, it was just great to be able to bounce all kinds of good and bad ideas off Walter; in this still heated-up topic, it is probably impossible to find anybody else, who is that knowledgeable but at the same time not drawn directly into the "frenzy".

Having developed a long friendship, I asked Walter to be my best man, when Christl and I got married in Santa Barbara in 1994. Walter agreed and in the shown picture you see him, the Scalapinos and our priest in the beautiful "sunken gardens" of the Santa Barbara courthouse.

Dear Walter, we feel joy and gratitude thinking back. Christl and I wish you all the best so that the "Kohn community" can enjoy your company and friendship for many years to come.

**About the Author**: Werner Hanke holds a chair at the University of Würzburg and is the dean of the department of physics and astronomy. His research is concerned with many-body physics in condensed matter. His current interests center around strong correlations and include developing a microscopic theory for high-$T_c$ superconductivity. Contact: Institute for Theoretical Physics and Astrophysics, University of Würzburg, Am Hubland, D-97074 Würzburg, Germany; hanke@physik.uni-wuerzburg.de; theorie.physik.uni-wuerzburg.de/TP1

# Thank You Walter for All You Have Taught Us

Michael J. Harrison

Michigan State University, East Lansing, U.S.A.

Walter and friends in front of ITP at the Walter Kohn 70th Birthday Symposium at UCSB

How many physicists do you know whose late-night reading includes 16th century French literature in the original? I know one; Walter Kohn! It was early in 1981, and my wife Ann and I had come to Santa Barbara, where Walter had kindly permitted me to occupy a desk during my sabbatical at the Institute for Theoretical Physics at UCSB. I had been doing administrative work as a dean at Michigan State University, and I very much wanted to do physics again and learn what were the interesting research problems that people were studying. Walter and Mara were most hospitable, and when Walter learned that Ann taught medieval and post-medieval French literature at MSU their conversation turned to Montaigne and his work. As it happened, Ann had a second copy of Montaigne's ESSAIS, and was quite happy to present the volumes to Walter as a gift. And we know that he read them in the weeks that followed.

Our year at UCSB was rewarding in large measure due to Walter's style of gentle leadership at ITP and his encouragement to me personally. I remember fondly the group trip to Anacapa Island that was arranged by him, and our conversations when we sometimes shared part of the bus ride from ITP back towards Santa Barbara at the end of the day. Walter's example of how to live a life in physics is truly something to be admired. It was great to return to UCSB for the celebration of his 70$^{th}$ birthday at a symposium in his honor, and I have enclosed a photo of Walter and attendees standing in front of Kohn Hall on that happy occasion.

It has been a rare privilege to know Walter Kohn. To a hundred and twenty!

**About the Author**: Michael J. Harrison is a professor in the Physics and Astronomy Department at Michigan State University. His current research in condensed matter theory concerns orbital magnetism and specific heat of quantum wires defined by 2-dimensional electron systems confined by parabolic quantum wells. He has also developed a statistical theory of frequency dependent auditory thresholds in humans and other primates. Contact: Department of Physics and Astronomy, Michigan State University, East Lansing, MI 48824-2320, U.S.A.; harrison@pa.msu.edu

# Vision Realized in Details:
# To Walter on His 80<sup>th</sup> Birthday

James B. Hartle

University of California, Santa Barbara, U.S.A.

The organizers of this well deserved volume marking your 80th birthday asked me to contribute a picture or anecdote concerning you. Alas, I have neither. However, I did not want this happy occasion to pass without expressing my appreciation for all that you have done for UCSB and for me personally. The organizers have graciously allowed me a brief note with which to do so.

I served a brief term as interim director of the ITP between Jim Langer and David Gross. You probably do not realize how much help you were to me in that. The style and principles with which you led the ITP as its first director were always before me. From you I learned from that every decision must have a reason and that every reason must be grounded in principles. I learned that one should consult widely, because other people have good ideas. I learned that vision is realized in details. I learned that one should be fair but purposeful, deliberative yet conclusive, and firm but flexible. Your example in these qualities was an inspiration. But perhaps most importantly, I learned from all those ITP dinners how to deliver a toast – a skill that has proved useful at the ITP and well beyond.

It is said that a wise man makes more opportunities than he finds. Thank you Walter for the opportunities you have created here, and the guidance your have given in realizing them. I hope that we have many more years of your example to follow.

**About the Author**: Jim Hartle is Professor of Physics at the University of California, Santa Barbara. His work is concerned with the application of general relativity to realistic astrophysical situations, especially cosmology. He has contributed usefully to the understanding of gravitational waves, relativistic stars, and black holes. He is currently interested in the earliest moments of the big bang where the subjects of quantum mechanics, quantum gravity, and cosmology overlap. He is a member of the US National Academy of Sciences, a fellow of the American Academy of Arts and Sciences, and is a past director of the Institute for Theoretical Physics in Santa

James B. Hartle

Barbara. Contact: Department of Physics, University of California, Santa Barbara, CA 93106-9530, U.S.A.; hartle@physics.ucsb.edu

# Walter Kohn in Japan

Hiroshi Hasegawa

Nihon University, Tokyo, Japan

Walter Kohn was known in early days in mathematical physics for his scattering theory when he was professor in the Carnegie Institute of Technology. At the time he had a Japanese friend, Tetsuro Inui, an applied mathematics professor in Tokyo who introduced me to him. In this period, he was actively working in solid-state theory (Kohn–Luttinger theory of one-electron motion in semiconductor). Moreover, he was becoming famous by the Kohn–Rostoker band calculation method, the Kohn anomalies and so on.

I was associated with Professor Kohn as a postdoc from August 1958 to July 1960 in the Carnegie Tech., and further on, one year in the University of California, San Diego (La Jolla). One day after my arrival at the Carnegie Tech., he gave me and my colleague, Bob Howard from Oxford, a problem theme on the dimensional electron density of states, i.e., electrons moving both in the action of a Coulomb impurity potential and a strong magnetic field; strong enough to form the Landau level first. It was a major subject studied experimentally in magneto-optics etc. His motivation was the question whether the one-dimensional density-of-state edge be observable as a peak of the optical excitation spectrum: the question was debated in a meeting of the American Physical Society where he raised the doubt, because he knew that under long-range Coulomb potential the near-zero energy scattering states are not as dense as for free electrons.

Bob Howard studied keenly the above topic by looking at a specialized applied mathematics book, Higher Transcendental Functions I, where a chapter is devoted to "confluent hyper-geometric functions" (on which T. Inui was an expert), communicating to me that the above suppression effect is really true. So, Bob proposed to write a paper to be submitted in the spring of 1960 when Professor Kohn had left already for California. But our writing the paper became delayed, unfinished by the time of my move to join him: there was dissatisfaction in our first treatment of the one-dimensional Coulomb scattering problem, i.e., a smooth connection of the outward and the inward solutions, which I had to convince myself of.

Finally in La Jolla, I got a satisfactory solution of the connection problem: an asymptotic outward solution and the small-variable expansion of the hyper-geometric function, both having a logarithmic term and the latter with an unknown parameter: the condition for a connection was found by equating both terms to yield, through determination of the parameter, all the bound-state energy eigenvalues in form of quantum defects for the Balmor series with a special ground state whose binding energy increases as $(\log B)^2$ in terms of the magnetic field strength $B$. I communicated this result from La Jolla to Bob Howard in Washington D.C. who got it published in 1961.

I met Walter three times in Japan: the first time was Autumn 1961 shortly after I returned from UCSD when he was invited by the Yukawa Institute for Theoretical Physics, Kyoto University. The second time was July 1965 when he participated the Ōiso International Summer School (Ōiso is a resort town near Tokyo) organized by Professor Ryogo Kubo. This time was a year after the paper of Hohenberg and Kohn had been published, the initial impetus of the density-functional theory. He was focussing his lecture on this work, while Professor Kubo (who died in 1995) gave a summary lecture on "fluctuation-dissipation theorem".

The third meeting in Japan was in 1990: he was invited by the Institute for Fundamental Chemistry (IFC)[1] which was founded in 1984 by Professor Ken'ichi Fukui (Nobel prizer in chemistry, 1981). This time, Walter took Mara with him: for a period of eleven days, they visited several institutions including the Solid-State Physics in Tokyo and the Molecular Sciences in Okazaki before coming to Kyoto. In May 25, I attended the sixth lecture session at IFC, organized by the director Fukui (who regretfully died in 1998). It started with an introductory talk by Professor Parr, a trustee member of IFC, who ended his talk by saying "Expert in mathematical physics, collision theory, solid state physics, and surface physics; distinguished chemical physicist; I am pleased to introduce Walter Kohn" (from the proceedings of IFC, 1990).

After all the formal were over on Sunday, May 27th, they went to see "Noh plays" at Kongoh Nohgakudo in down-town of Kyoto, and my wife and I joined them. The major program of the plays was "Tomonaga" (a tragic story about the death of the eldest son Tomonaga of Yoshitomo Minamoto); one of the three Warrier Plays of the Middle Ages in Japan. It consisted of two stages; the 1st stage a scene of a woman asking a priest to perform a requiem mass for sixteen-years old Tomonaga at his gravesite, and the second the appearance of Tomonaga's spirit recalling the events that led to his death. It was of particular interest to see the main actor (*shite*) play two entirely different roles, i.e., the old woman and the young Tomonaga. The player was Iwao Kongoh, Grand Master of the Kongoh School. In the evening, we invited them for supper at our home place. I took this occasion to tell him a bit about my studies.

---

[1] Now, Fukui Institute for Fundamental Chemistry, Kyoto University, Sakyo-ku Kyoto City.

*Left*: Robert G. Parr giving an introductory talk on Walter Kohn; *Right*: Walter Kohn giving a lecture on *density-functional theory* at the sixth lecture session of IFC, 1990 (photo: IFC proceedings 1990)

Prior to their visit to Japan, we stayed in Santa Barbara for three months in order to participate in a long-term research project in 1989 at the Institute for Theoretical Physics, UCSB, to study quantum chaos; one of the world-wide trends in physics, or better to say, in dynamical theory: my motivation to join this field was an extended version of the very theme that Professor Kohn gave me in 1958, namely, the subject of the "hydrogen atom in a magnetic field", a typical non-integrating dynamical system with degree of freedom 2. But what I wanted to communicate to him was the paper of 1961 with Howard. Indeed, after 1977 the quantum theory of the hydrogen atom in a strong magnetic field was taken up by J. Avron, I. Herbst, and B. Simon as an important subject in mathematical physics (functional-analytic studies of Schrödinger operators) whose result was compared with ours (Commun. Math. Phys. **79**, 529 (1981)). I demonstrated it at the Como Conference on Fundamental Aspect in Quantum Theory held in 1985.

Walter never talked to me about politics, or his opinions about world politics, except the last time I met him when I visited UCSB in late August, 1990: at the municipal airport, he told me of his enthusiasm for the monumental speech entitled "Der 8. Mai 1945 – 40 Jahre danach" given by Richard von Weizsäcker, brother of Carl F. von Weizsäcker a physicist (student of W. Heisenberg) and the former president of Germany at the time of its reunification.

**About the Author**: Hiroshi Hasegawa, a senior researcher at the Institute of Quantum Science, Nihon University. He is undertaking researches in the field of quantum information theory and quantum chaos: specifically, application of random matrix theory to metal-insulator transition. Contact: IQS, Nihon University, Kanda-Surugadai, Chiyoda-ku, 101-8308, Tokyo, Japan; h-hase@mxj.mesh.ne.jp

# My Favorite "Walter Story"

Alan J. Heeger

University of California, Santa Barbara, U.S.A.

Shortly after the announcement in October, 1998, that Walter Kohn had been awarded the Nobel Prize in Chemistry, our newly appointed Executive Vice Chancellor arrived on the UCSB campus. The Physics Department invited the EVC to visit the Department and to meet our distinguished faculty members. After introducing the EVC to the Physics faculty, the Department Chair asked that we introduce ourselves individually and one at a time in turn around the room, and that we each say few words about our research interests. As Professors will often do, the comments by many of the faculty were a little drawn out. When it came to Walter's turn, the great physicist said simply the following (with a characteristic smile on his face): "My name is Walter Kohn and I do Chemistry".

As a physicist, as a scientist, as a humanitarian, as a colleague, as a friend – and even as a chemist, Walter Kohn is an inspiration.

Walter, we love you. Happy 80$^{\text{th}}$ Birthday!

**About the Author**: Alan J. Heeger serves as Professor of Physics and Professor of Materials at the University of California, Santa Barbara and also heads a research group at the university's Center for Polymers and Organic Solids. He was awarded the Nobel Prize in Chemistry (2000) for his pioneering research in and the co-founding of the field of semiconducting and metallic polymers; his research efforts continue to focus on the science and technology of semiconducting and metallic polymers. Current interests include studies of conjugated polyelectrolytes, and the use of such luminescent water-soluble semiconducting polymers (and oligomers) as components in bio-specific sensors. Contact: Center for Polymers and Organic Solids, University of California, Santa Barbara, CA 93106-5090, U.S.A.; ajh@physics.ucsb.edu; www.cpos.ucsb.edu

# Walter Kohn as a Scientist and a Citizen

Conyers Herring

Stanford University, Stanford, U.S.A.

For over half a century I have enjoyed speaking and corresponding with Walter Kohn, and studying some of his work, and over this time I have felt a growing respect for the depth of his insights and for his conscientiousness as a member of the scientific community.

A couple of hasty limericks come to mind:

> In some of the papers by Kohn
> (with colleagues, or often alone)
> the ideas he discloses,
> like the laws brought by Moses,
> deserve to be carved into stone.

> Just like luring a dog with a bone
> octogenarian Kohn
> sics his pet named "equations"
> against false ideations
> and when the job's done, truth is known.

**About the Author**: Born in 1914, Conyers Herring developed a boyhood interest in astronomy and physics. In graduate school he eventually switched his field of specialization from astrophysics to solid-state theory, and obtained a Ph.D. from Princeton University in 1937. After a few post-doctoral years he became a faculty member at the University of Missouri, but in the great mobilization of physicists for war work he had to drop physics for operational research on anti-submarine warfare. After the war he became a professor at the University of Texas, but soon was recruited to join, in 1946, the solid-state physicists that Bell Telephone Laboratories was energetically recruiting. There he stayed until retiring to become Professor

Conyers Herring

of Applied Physics (now Emeritus) at Stanford University. He is a member of the National Academy of Sciences, and has received a number of prizes: Oliver E. Buckley Solid-State Physics Prize, American Physical Society (1959); J. Murray Luck Award for Excellence in Scientific Reviewing, National Academy of Sciences (1980); von Hippel Award, Materials Research Society (1980); Wolf Prize in Physics, Wolf Foundation (1985). Contact: Geballe Laboratory for Advanced Materials, MC 4045, Stanford, CA 94305-4045, U.S.A.

# A Personal Tribute to Walter Kohn on His 80<sup>th</sup> Birthday

Pierre C. Hohenberg

Yale University, New Haven, U.S.A.

I first met Walter Kohn in 1963 when I was a fresh PhD spending a post-doc year in Paris at the Ecole Normale Superieure in the group of Philippe Nozieres. As it happened, Walter was also spending time at the Ecole Normale and conditions being what they were I was privileged to share an office with Walter during an extended period. If I remember correctly, this relatively large space was Philippe's own office and even if it wasn't I remember it to have been a general meeting place and thoroughfare, a little bit like trying to think deep thoughts in the middle of Times Square.

In any event, soon after I met Walter he did me the honor of inviting me to join him in a new research project he was undertaking, the examination of theoretical methods for treating the inhomogeneous electron gas. To say that this project turned out to be successful is a singular understatement, since the paper we produced in the Spring of 1964 [P. Hohenberg and W. Kohn, *Phys. Rev.* **136**, B864 (1964)] was one of two works cited by the Nobel Committee in awarding the 1998 Chemistry Prize to Walter. At the time I remember that it felt like being led through an enchanted, and at times haunted, forest by a trusted guide. The journey was both instructive and entertaining, but I have some trouble identifying what I could possibly have contributed that would merit Walter's generosity in placing my name first on the 1964 paper. One incident stands out in my memory. After we had proved the primary theorem on density functionals we started exploring possible applications to real systems. When Walter suggested using the theory to make improvements in the prevailing methods for determining the band structure of solids, I said to him that I feared this would require expertise in practical areas of materials science I knew nothing about. He then drew himself up to his full five feet ten inches and said to me, "Young man, I am the Kohn of Kohn and Rostoker!" [*Phys. Rev.* **94**, 1111 (1954)].

The year 1963 was a difficult one for me personally since in November, while I was at the Ecole Normale, my father died suddenly in Paris, where my parents lived. Like Walter, my father was born and grew up in Vienna, but he had emigrated to France in the nineteen twenties and so was spared the

The Kohn and Hohenberg families getting ready to attend the Nobel Prize ceremony in Stockholm in December 1999. From left to right, Pierre Hohenberg, Mara Kohn, Barbara Hohenberg, Walter Kohn (photo: P. Hohenberg)

trauma of leaving his home town when Hitler marched in after the "Anschluss" (1938). The Viennese connection is another link I feel with Walter since he represents a kind of intermediate step for me between my father, who was an adolescent during World War I, and my own experience growing up during and after World War II.

During my year in Paris I discussed with Walter my search for a position the following year and sought his advice. Having gone through the standard list of university departments Walter asked me "Have you thought about Bell Labs?". I answered "What's that?". He then told me about the Bell Labs theory group where he and Quin Luttinger had spent so many fruitful summers, and concluded "I think you should definitely apply there. Spending a year or two at Bell Labs will do you a lot of good, it will professionalize you".

As it turned out I followed Walter's advice and went there for one more postdoc, which turned into a job lasting 30.5 years to the day. To my own surprise I found Bell Labs to be the right mix of "professionalism" and scientific freedom and I am grateful to Walter for having steered me to that great institution.

My scientific interests did not remain with electronic structure so I watched the development of Density Functional Theory as an admiring outsider, observing how Walter steered it forward with persistence and vision. An incident I remember clearly occurred when I saw Walter some time in the mid-eighties and he was reporting to me on the progress of "our baby". He said to me "Things are heating up. The chemists have finally caught on". Prophetic words.

After Paris I next came into close contact with Walter in 1980 when he invited Jim Langer and me to organize a research program at the newly opened Institute for Theoretical Physics in Santa Barbara, whose director he had just become. Jim and I accepted his invitation and "Pattern Formation in Nonequilibrium Systems" became the first full-fledged research program of the new institute, attracting scientists from many centers throughout the world and from many fields inside and outside of physics. For Jim Langer, who had been a colleague of Walter's at "Carnegie Tech" early in his career, this visit presaged a move to Santa Barbara, where he eventually became the director of the Institute. For me the year spent in Santa Barbara marked a shift in scientific interest to the field of nonequilibrium physics and pattern formation, an interest that has persisted for many years. My connection to the Institute also remains to this day. I have visited many times, participated in conferences and research programs and was a member and chair of its Scientific Advisory Board.

The wonderful news of Walter's Nobel Prize held special significance for Lu Sham and me since we were his junior collaborators and co-authors on the two seminal papers cited by the Nobel Committee. Walter did us both the honor of inviting us to join him and his family in Stockholm for the prize ceremony, along with Bob Parr (University of North Carolina) a long-time friend and colleague. Unfortunately, health problems prevented Lu from attending and he was sorely missed. The pageantry of this greatest of all scientific rituals and the opportunity to participate publicly in honoring a mentor and friend were truly unforgettable experiences for which I will always be grateful to Walter.

Some years ago I decided on a drastic change of professional orientation, abandoning the quiet life of a research scientist for that of a university administrator. In preparing for this change I assembled a list of personal references and I asked Walter if he would be willing to write on my behalf, since he had knowledge of my organizational abilities from my work at the ITP. Walter's immediate reaction was "Don't do it! I've seen better men than you trampled underfoot and completely destroyed by university administration". In this one case I did not heed Walter's advice; I did not ignore it, I just decided otherwise. After nearly eight years as Deputy Provost for Science and Technology at Yale University I will not say that I regret my decision, but I will affirm that I understand more fully what Walter was talking about! As I now contemplate re-entering the world of science and facing the prospect of formal retirement I look to Walter as a shining example of intellectual vigor and general wisdom. I eagerly await his next piece of good advice.

**About the Author**: Pierre Hohenberg is currently the Deputy Provost for Science and Technology at Yale University, where he oversees the physical, engineering and biological sciences. He also acts as the chief research officer for the university. His

Pierre C. Hohenberg

research has ranged over electronic structure, theory of superconductivity and super-fluidity, the study of critical phenomena and phase transitions (especially dynamics), and pattern formation and chaos in spatially extended nonequilibrium systems. Hohenberg spent the bulk of his research career at Bell Laboratories, from which he departed in 1995 to take up his current position. Contact: Departmet of Applied Physics, Yale University, New Haven, CT 06520, U.S.A.; pierre.hohenberg@yale.edu

# Recollections of a Long Friendship

Daniel Hone

KITP, University of California, Santa Barbara, U.S.A.

I think the first time Donna and I met Walter was in Paris the summer of 1963. I was finishing a postdoc there at the Ecole Normale Supérieure, and he was coming for a sabbatical year's stay. Shortly thereafter we left Paris, in a display of spectacularly bad timing, missing the excitement of that year's first work on density functional theory, which ultimately led to Walter's Nobel Prize. But I do want to take credit for an important contribution we made to that work. We turned over our Paris rental apartment to Pierre Hohenberg, who was the one who wrote the seminal first paper on density functional theory with Walter that year.

Sometimes patience and persistence pay off. Some 35 years later I had the chance to begin a scientific collaboration with Walter. It has been a wonderful experience. About four years ago we realized that we had independently become interested in problems connected with systems driven by strong time periodic fields, such as lasers. Reflecting a lifelong pattern of enjoying the interactions with junior scientists, Walter brought in as a co-worker with us on this project a young researcher from Germany then visiting the ITP, Roland Ketzmerick. Extensions of that work continue to this day between the three of us, and it has brought me, among other things, an exposure to the intercontinental travel for scientific collaboration that has been so much a part of Walter's life. Most importantly, from a scientific point of view, it has given me a personal insight into Walter's deep physical understanding, and his unusual insistence on understanding absolutely every implication of speculations and assertions made in connection with the research.

In this collaboration I have seen another revealing aspect of Walter's unusual abilities. Because of the vital way in which physics research builds on the work of the past, it is important to be able to remember or find earlier results which will inform our current investigation. For those of us who find it hard to recall what we wrote or said a month ago, it is remarkable to watch Walter reconstruct a detailed calculation done 30 or 40 years ago. But even more amazing is to listen to him explain, as well, the **wrong** paths he and his colleagues took and rejected at the time. So when you hear Walter tell an anecdote from the similarly distant past, as we often have, you can trust to its accuracy; the quotes are likely to be close to verbatim. And it is not only his own work that is ingrained in Walter's mental filing system. Some years ago he instituted a weekly lunchtime gathering, which can include informal talks on almost any subject in the general areas of theoretical condensed matter physics and chemistry. It is rare that there isn't some point in the discussion, no matter what the subject, in which he doesn't recall some paper by himself or by others that sheds important insights into the work at hand, including the journal location and approximate date of the paper.

We were all delighted when Walter agreed to act as the founding Director of the ITP – now the KITP. The subsequent great success of the Institute speaks for itself as to the wisdom of that critical choice of first Director. Not everyone reading this may be aware that Walter agreed to accept this position not for the proposed period of five years, but only on condition that he could in fact return to UC San Diego after a maximum of two years. So it was a wonderful surprise when he asked after a year if we might be interested in him staying in Santa Barbara indefinitely. Of course, we leaped at the opportunity, and he has been here as a valued colleague and friend for more than 20 years. I have benefited personally, as well, in my position as Deputy Director of the KITP for the past 8 years. I continue to see the impact on the ongoing success of this unique Institute of the organization and administrative structure established by him so wisely at the beginning.

Over the last couple of decades we have frequently enjoyed, as well, the hospitality of Walter and Mara. At dinners and parties we have not only seen many mutual scientific friends, but have met a most interesting group of people from many other areas, reflecting their wide acquaintance and involvement with so many cultural, intellectual, and community activities. During recent years Mara was frequently in New York and elsewhere, but that seemed not to slow Walter's entertainment schedule, at all. He often would organize events at his house for groups of visitors, putting together his own culinary specialties for refreshments, notably taking advantage of the ubiquitous avocado in Santa Barbara to make his unique brand of guacamole.

Walter's interests are so incredibly broad that each of us is bound to find a big overlap with whatever our own might be. Some years ago Donna and I drove him from Santa Barbara to Lake Arrowhead, where a group of condensed matter theorists from the various campuses of the University of California campuses and families were getting together for what was then an annual scientific and social two day meeting. During the car trip of several hours the conversation ranged from family planning and the activities of Planned Parenthood, to population issues more generally, immigration, privacy questions and proposed personal identification documents, the ACLU, politics, performing arts, and other things I have surely by now forgotten. We have a large collection of Asian art, and he and Mara share an interest in that, partly through the large collection amassed by Mara's father. Seeing him at the local chamber orchestra concert reminded us of the love of classical music we share. We often talk about Paris, where we have both spent so much time and which we enjoy so much. But others certainly find a different but equally rich range of interests to share, since the breadth and depth of his own involvement and experience has been so great.

On a recent trip to Europe, in part to continue work with Ketzmerick in Göttingen on our project with Walter, we discovered that the Kohns were in Berlin. We were delighted to accept their invitation to join them for a day on a boat trip through Berlin organized by the Fritz-Haber-Institut, where he was a visiting. Given Walter's travel schedule it is almost unsurprising to find him nearby, wherever one might be at any given moment. It was wonderful to share the day with them, a recent event in a friendship which has lasted for many years and which, we hope, will continue for many years to come.

Boat tour on the river Spree in summer 2002. From left to right: Walter Kohn, Donna Hone, and Mara Kohn (front); Catherine Stampfl and Weixue Li (back) (photo: D. Hone)

**About the Author**: Daniel Hone is Deputy Director of the Kavli Institute for Theoretical Physics and Research Professor of Physics at the University of California, Santa Barbara. His research has covered many areas of condensed matter theory, including transport properties of liquid Helium-3, magnetism and magnetic resonance – often in low dimensional and impure materials, colloids and polymers, and the behavior of explicitly time dependent quantum mechanical systems. Contact: KITP, UCSB, Santa Barbara, CA 93106-4030, U.S.A.; hone@kitp.ucsb.edu

# My Recollections of Professor Walter Kohn

Naoki Iwamoto

Kagawa University, Takamatsu, Japan

Twenty two years ago I received a phone call from Walter Kohn. "I wish to offer you a postdoctoral position. Are you interested?" My answer was an immediate "Yes". I had finished my degree at Illinois, had seen enough of corn fields and was ready to see some mountains and ocean. A year later I arrived at the Institute for Theoretical Physics then located on the sixth floor of the Ellison Hall at UCSB. The offices were about to become full. I kept a stack of several boxes of books and photocopied papers in my office. Walter Kohn once said of them, "They look like a radiation shield."

I knew of the Hohenberg–Kohn paper although I had not worked on the subject. The paper is short, elegant and mysterious. The consequences appeared to be far-reaching. At that time I thought that the greatness of the Nobel Prize consisted not in the quality of the physicists who had received one but in the number of excellent physicists who had not yet received one.

In one seminar Walter Kohn asked the speaker a question, "What does the wave function look like?" He often seemed to be thinking of physics problems in terms of the wave function. Sin'itiro Tomonaga once said that there are not many physicists who really understand quantum mechanics. I believe that Walter Kohn is one of those few.

I also heard the following story. A speaker who gave a seminar talk didn't answer Walter Kohn's question well. Walter Kohn demanded a satisfactory answer, saying, "This is a very important point!"

In physics he accepted no excuses from his graduate students. One of his graduate students started to explain why he could not finish some calculations or whatever he was told to do. Shaking his head, Walter Kohn interrupted this student and said gently but firmly, "Just do it!"

His management style as director of the Institute was rather democratic. He would often call for an Institute member meeting including everyone and ask our opinions about how the Institute should be run and should be doing. As founding director, he was always thinking about the Institute. His sense of responsibility for this job was so strong that even after his by-pass heart surgery he was already back in his office only a few days later.

Walter Kohn holding a state of the art computational device (an abacus!) "You need one of these to apply the density functional theory to real problems." At his 60th Birthday Party held at ITP, UCSB, 13:30–14:30 on Wednesday, 9 March 1983

There were many social activities at the Institute, in which he participated actively. Hiking in the near-by mountains and picnics were held often. He also invited us postdocs to his house in Santa Barbara once in a while. He would ask, "Are you free this Saturday? How about coming over to my house for dinner?" On one such occasion, his wife was out of town. So he cooked meat in the oven, prepared and served dinner for us all by himself. His hospitality of this kind is still fresh in my memory. He really cared about the members of the Institute.

Walter Kohn has deep appreciation of music and art. In the rooms and hallways at the Institute several modern paintings that he chose by himself were hung on the wall. It was not until recently that I had a chance to talk with him about art, music and photography. From the Office of the Dean of the Graduate School at The University of Toledo, I was asked to call him on some business. In the previous few month the Toledo Museum of Art had held an exhibition of Picasso's print works – a collection from Norton Simon Museum – (which I liked so much that I visited twelve times) and an exhibition of Philippe Halsman's photography. There was also Yo Yo Ma's concert at Bowling Green State University. When I was asked, "How is life there? Are you happy?" I started to talk about these exciting things and Walter Kohn made relevant comments on each of them. We had a wonderful conversation over the phone for quite a long time. I am sure that he would enjoy Isamu Noguchi Garden Museum (Noguchi's former studio) in the suburbs of Takamatsu.

He was always gentle, smiling and seemed to know everything. To me he has been a father-like figure. In fact my father turns 80 in March 2003, too!

80 sai no otanjoubi omedetou gozai masu.

**About the Author**: Naoki Iwamoto is Professor of Physics at Department of Advanced Materials Science, Kagawa University, Japan. His research interests are in high-energy astrophysics, condensed-matter physics and plasma physics. He has published single-authored papers in Physical Review A, B, D, E, and Letters. Contact: 2217-20 Hayashi-cho, Takamatsu 761-0396, Japan; niw@eng.kagawa-u.ac.jp

# With Walter: 40 Years of Friendship

Denis Jérome

Université Paris-Sud, Orsay, France

In the Fall of 1963, Walter was at Ecole Normale visiting the lab of Philippe Noziéres. One day he came to Saclay invited by Anatole Abragam. At that time I was a student working for my PhD thesis in Abragam's lab and since I was involved in a project of metal-insulator transition in doped silicon it was quite natural that I meet Walter who had published some years earlier his basic article with J.M. Luttinger on the physics of impurities in semiconductors. This was my first encounter with you Walter.

You remember probably Walter that we talked about the Mott transition which I was studying in doped silicon but we also discussed the possibility for you to join us for a day out in the Fontainebleau forest. Every sunday, a group of friends at Saclay used to climb small rocks in the forest. You came one sunday with the whole family including Rosalyn who was still a young baby. I suppose you thought that we were rather strange climbing up and down these tiny rocks with tremendous effort. Maybe that reminded you similar practices in the Austrian alps?

I was so impressed by your kindness during this first meeting let alone your knowlege in physics which I was not yet fully able to appreciate that I cherished the possibility of spending one year as a post doc in your group. A. Abragam who was my thesis supervisor wrote you a letter and to my great satisfaction and surprise you answered in a positive way.

This answer was not obvious for me since I was an experimentalist of no help for you already head of a renowned theory group at La Jolla. After a very enjoyable crossing over the Atlantic on board the France and three weeks driving from coast to coast I finally reached La Jolla in November 1965 very excited by this long trip and also worrying about how I would establish in the Far West. Thanks to you and Lois I received a very warm welcome and I got to meet quickly the friendly community around the University. I remember the very first day when we met in your office you told me to call you Walter. In France we were still before 1968!

I kept a very warm memory of the numerous invitations to your home, especially the Seder in which I took part. In terms of research I was teamed

Denis Jérome

December 2001: Walter with Mary Friedel and Sir Sam Edwards (*left*); Dinner in Paris. Walter facing Jacques Friedel, Mara, and Peter Fulde (*right*)

with Maurice Rice on the problem of the excitonic instability in which you were deeply involved. Besides the impressive weekly meeting the three of us in your office I was enchanted by the delicious smell of your pipe.

I owe you an introduction to high pressure physics since we had found out that one possibility to observe the excitonic phases was ytterbium under pressure. This early introduction became decisive later since once back to France J. Friedel thought I had become a specialist in high pressure and asked me to start in 1967 a new group in his lab studying the physics of metals and alloys under high presure and low temperature. Even if we failed to discover the excitonic insulator in ytterbium my interest in that kind of physics has been crucial later for the field of organic superconductors.

We had become very close friends and in September 1967 as you were spending some more time in Paris I trusted you to be the witness when I married Vered in the small village of Jouy en Josas. I never had to regret both choices! Later we had the opportunity to meet when you came almost every year for your immersion into the French civilisation. Your practice of French had improved so quickly that I realized that you were becoming more fluent in French than I was in English. So, let me finish in French.

Nous avions toujours des rencontres sympathiques soit chez nous soit chez nous à Jouy soit chez Chuck et Michelle rue du Sommerard, agrémentées d'un bon repas copieux et bien arrosé. Plus récemment, nous avons été très touchés que tu acceptes malgré un emploi du temps chargé de venir à Paris célébrer les 80 ans d'un ami commun qui nous est cher, Jacques Friedel. Ce fût l'occasion d'un dîner avec de nombreux amis et en particulier avec la présence de Mara. Nous avons aussi pu écouter la brillante allocution que tu as donnée à l'Académie des Sciences, au cours de laquelle tu nous a remis en mémoire les liens étroits d'amitiés qui existent entre toi et la France (voir photo).

I chose to write these few sentences in French because one always feels and you confirmed it that France and French are your second home country.

**About the Author**: Denis Jé
oratoire de Physique des Solides, Orsay. He is concerned with experimental Condensed Matter Research with emphasis on high pressure physics. He has contibuted to the discovery of superconductivity in organic matter and is currently involved in the study of strongly correlated one and two dimensional conductors. Contact: Laboratoire de Physique des Solides, Université Paris-Sud, F-91405 Orsay, France; jerome@lps.u-psud.fr

# The Noble Day

Alex Kamenev

University of Minnesota, Minneapolis, U.S.A.

I consider myself to be a really lucky guy for being one of Walter's post-docs. That was amazing three years, with almost daily meetings with Walter where long, sometimes exhaustive, search for physics elegancy was mixed with endless stories about famous (and not) physicists, Shakespeare poetry, campus architecture, French cuisine and what not.

That day in the fall of 1998 started as any other with me preparing a breakfast and CNN making little sense in the background.

Suddenly something attracted my attention: "...in chemistry was awarded to John Pople from Northwestern University and Walter Kohn from the University of California ...". What?!?! Walter, chemistry ...what is it all about? I neglected my coffee and rushed to the internet. No mistake: tonight it was announced that Walter won the Nobel prize in chemistry. I was so excited that even risked to wake up my wife:

– Listen, listen, Walter got a Nobel prize! What should I do?

– Get the best flowers you can find in this city and go to his home.

I followed the advice, but decided to pass through the Institute to pick up Ann Mattsson – the other post-doc of Walter. In the ITP Deborah Storm was already in the midst of preparations for the Big event. The holy duty of all Institute post-docs is to move heavy furniture every time some event is approaching. After this refreshing experience we were finally on our way down to southern Santa Barbara, where Walter lives. With 85 miles per hour we could be an easy prey for cops, but that was our lucky day. At 9 a.m. we are already at La Vista Grande. The familiar sticker on the bumper: "Population, God stopped on two, what about you?" OK, Walter is still at home. A modest guy with a beard approaching us while we are parking:

– Are you students of Professor Kohn?

– Sort of.

– I am from "Los Angeles Times", I drove all the way from LA to talk to the professor. Would you mind, if I enter with you?

– Well, we don't, but this is not our house ...

And finally here is Walter! He radiates happiness with all his look. Journalist with a beard jumps ahead with his most precious question:

– How do you feel, Professor?

– I am feeling great!

And it is so obvious without a word. Telephone rings. Mara, trying to maintain calm and order, welcomes us into the house. In 10 minutes Walter is back from the phone. We: "Walter, we are so ...", Journalist: "Were you expecting it?", Walter: "No."

Telephone rings again. Walter is gone for another 10 minutes. Meanwhile Mara is telling us how they were woken up in the middle of the night by the phone call: "We are terribly sorry for bothering you, this is the Royal Swedish Academy of Science ...".

Walter is back, but the door ring is ringing. This is a camera crew from the Nobel committee headed by a very young and very arrogant director. They want to produce a short movie clip about Walter. The director is obviously at home, he expels the LA journalist and asks to remove the chandelier from the sealing. Mara is upset and tries to minimize the damage. Walter, Ann and myself are sit around the table with papers deliberately spread all over it and ordered to discuss some chemical (???) problems.

We look at each other and can not help smiling. "No, no, no this is not good. You must be serious. Chemistry (!!!) is a serious science. Once again please." Finally we open a draft of our paper and try to discuss some changes. Walter does his best to suppress the smile and look seriously, so do we. Telephone keeps ringing without a break. Finally the comedy is over. The director is satisfied with our chemistry.

The movie people asked me to show them a way to the Kohn hall (Walter once told me: "I did not have to die to have my name on a building, but the most amazing is that I did not have to pay for it either."). We all left, Walter and Mara followed shortly after. In the Institute everything was already prepared for the press conference. The big hall was full of people: University colleges, press, friends, general public. After congratulations and applause and more congratulations Walter told the story of his life. He was obviously touched and agitated by the events of the day. He spoke for more than an hour. I especially remember the words of admiration towards his Ph.D. adviser Julian Schwinger. Walter also told about his childhood in Austria and the role the Austrians played in the Holocaust and in the life of his family. This part was reproduced by many news agencies. The next day Walter received a telegram from the Austrian president with congratulations, but no acknowledgment to the harsh words that were said a day before. Walter was very upset and drafted a polite but strict reply. As far as I know, there were several rounds of letter exchange between them that found their way to the major Austrian newspapers. I remember, Walter, receiving e-mails of support

Alex Kamenev

from Austrian physicists for his role in initiating discussion on the country war history.

After the press conference the comedy started all over again. The movie people wanted to picture Walter in front of the big blackboard discussing science. I was hired to cover the blackboard with "some important equations professor Kohn is currently working on". I wrote a mixture of Schrödinger, Poisson, continuity, etc, equations – the one I was reasonably capable to write without mistakes. You can still see them on the official poster of the Nobel committee dedicated to the 1998 prize in Chemistry. Walter is pictured there in front of the blackboard covered with all these "important equations". I take a full responsibility for any mistake you may spot there. After that Walter had to imitate a scientific discussion once again.

Everybody was a little tired for jokes and that probably helped to satisfy the young director's idea about the "true scientific process".

That was one great day, one of the three great years.

**About the Author**: Ph.D. Solid-State Physics, Institute of Science, Rehovot, Israel, 1996; M.Sci. Theoretical Physics, Moscow State University, 1987; Faculty, University of Minnesota, 2001–present; Assistant Professor, Physics Department, Technion, Israel Institute of Technology, 1999–2001; Postdoctoral Researcher, Physics Department and Institute for Theoretical Physics, University of California at Santa Barbara, 1996–1999; Junior Researcher, Institute of Radioengineering and Electronics, Academy of Science, USSR, 1987–1991; Research Area: Theoretical condensed matter physics, disordered systems and glasses, field-theoretical treatment of many-body systems, mesoscopic systems, out of equilibrium systems. Contact: Department of Physics, University of Minnesota, 116 Church Street S.E., Minneapolis, MN 55455, U.S.A.; kamenev@physics.umn.edu

# The Pleasure of Getting to Know Walter Kohn

Boris Kayser

Fermi National Accelerator Laboratory, Batavia, U.S.A.

I had the wonderful luck of getting to know Walter Kohn when he was the first Director of the Institute for Theoretical Physics in Santa Barbara. At the time, I was at the National Science Foundation, and both we and the folks in Santa Barbara were trying to launch this new adventure in nurturing theoretical physics in the best possible way. As I soon learned, Walter is a man of very considerable old-world charm and erudition, and he is such an exceptionally nice and supportive person that I came to think of him as "Uncle Walter". As Director of the Institute, he went to great lengths to create a welcoming and nurturing environment for all the participants and visitors. In this he was extremely thoughtful. An example: Walter realized that when deciding which people to invite to the Institute as postdocs, one should take into consideration what scientific programs will be running there during the time that a candidate postdoc would be present, and ask whether this collection of programs would make a strong and helpful scientific environment for this postdoc.

This book contains many stories from contributors who know Walter as the terrific scientist that he is. I came to know another side of Walter – the good-natured, personable man who, with great effectiveness, led the Institute in its formative years, eager to promote physics and to foster the scientific careers of those who crossed his path.

The time I am writing about was long ago (about 20 years), but just thinking about the Walter Kohn I came to know then brings a great big smile to my face.

**About the Author**: Boris Kayser is an Elementary Particle Theorist. For 1972 to 2001 he worked at the US National Science Foundation, and since 2001 he is at the Fermi National Accelerator Laboratory. Contact: Fermilab, P.O. Box 500, Batavia, IL 60510-0500, U.S.A.; boris@fnal.gov

# Happy Birthday Walter!

Roland Ketzmerick

Technische Universität Dresden, Dresden, Germany

I got to know Walter in the summer of 1992 in Santa Barbara when I began a post-doc year with him. Coming from a different field (quantum chaos) I knew only one or two of his papers and was pushed by my "Doktorvater" Theo Geisel to be a post-doc of Walter, while my wife Astrid and I were mainly looking forward to spending a year at the California Coast. I did not have the slightest idea what a fortunate man I was!

Walter Kohn in his ocean view office at UCSB in the summer of 1993 with his post-docs Yigal Meir (*right*) and Roland Ketzmerick (*left*)

For anyone who knows Walter it is not a surprise to hear how much I benefited from working with him – not only scientifically! His openness, hospitality, interest and knowledge about almost everything is so remarkable. Scientifically I profited a lot from being around him and going to his weekly lunch meetings, but still it took almost a year until we found a project where I could reasonably contribute (Floquet systems). This was the starting point of an ongoing collaboration (together with Dan Hone from the Institute for Theoretical Physics in Santa Barbara) which took place either during my subsequent visits to Santa Barbara or Walter's visits to Frankfurt, Gö very recently Dresden.

Writing a paper with Walter is a long struggle: First one has to get a project to a point of complete understanding from many perspectives. Only then he would start thinking about writing it down – in contrast to the rapid publication sequence of partly understood results, so common nowadays. But then he is determined to get it done: Once, we spent 5 hours on a beautiful sunny Sunday in the office in order to get the paper ahead – by about two paragraphs. On that day, however, I learned quite a bit about writing scientific papers.

Walter's 70$^{th}$ birthday was during my post-doc time with him. So Yigal Meir (the other post-doc at the time) and I bought some wine and cake for a little gathering. When Walter turned 75 he gave a big dinner party at the Institute for Theoretical Physics. Among the many speeches one was given by the chancellor of UCSB, Henry Yang, who proclaimed that if there were a combined Nobel prize for physics and peace Walter would certainly get it. I am absolutely sure, at that evening nobody thought about chemistry.

We always enjoyed being guests at Walter's and Mara's house. Recently, our kids had a memorable afternoon at "La Vista Grande". Carina and Erik were fascinated by the orange trees in the garden. They asked Walter if they could pick some oranges. He kindly walked with them around the house and they picked quite a few. Mara, entertaining the kids all afternoon, invited them to the kitchen and together they made some fresh orange juice which our kids were drinking happily. Whenever they press orange juice now (unfortunately, from oranges bought in the store) they remember that afternoon.

What a fortunate start I also had in my new hometown Dresden: The first guest we invited to our home was Walter and the second was Yigal Meir, my office-mate during the post-doc year with Walter.

Walter, thanks for being such a good friend over many years.
Happy Birthday!

**About the Author**: Roland Ketzmerick is a professor of Computational Physics at the Technische Universität Dresden. He is interested in quantum chaos, i.e., the understanding of and the search for quantum signatures of classically chaotic dynamics. In particular he enjoys mixed systems, where classically regular and chaotic behavior are intertwined in a fascinating way. Contact: Institut für Theoretische Physik, TU Dresden, D-01062 Dresden, Germany; ketzmerick@physik.tu-dresden.de; www.comp-phys.tu-dresden.de

# Memories of a Great Scientist and a Great Person

Barry M. Klein

University of California, Davis, U.S.A.

I first met Walter Kohn in person in 1986 when I visited him at the University of California at Santa Barbara in my capacity as a Program Manager at the NSF where I was on a one-year temporary appointment on a leave from the Naval Research Laboratory. Of course, I "met" Walter a lot earlier in my studies as a graduate student and in my professional career research work and readings and writings that followed.

I recall our first meeting at UCSB in some detail because Walter tends to leave a unique impression on you – kind and caring, a wonderful teacher in the generalized sense, and someone so very much engaged in communicating to you his joy and love of physics (and chemistry!) and his specific research. He wants you to understand! I was certainly more nervous as the NSF reviewer rather than he was as the reviewed, so I just concentrated on enjoying the moments. We had a wonderful conversation about his work.

When I joined the University of California at Davis in 1992, I became a formal colleague of Walter's, and I made several trips to the Institute for Theoretical Physics at UCSB that Walter founded. One trip was special for me when I gave a lecture at ITP that Walter attended. I talked about some of my electronic structure results and also wandered into discussing some fairly esoteric aspects of density functional theory that I prayed were up Walter's expository standards. Although the DFT details were not central to my presentation, I think that I deliberately wandered into that part of my talk just to be sure that I could engage Walter in a discussion about DFT. In his characteristic fashion, Walter was the teacher, not the critic, and I immensely enjoyed our ensuing dialogue.

Since joining the University of California, I also learned that Walter, besides being an intellectual giant in condensed matter theory, also has a reputation for being very engaged in University of California issues over the years, bringing the same skills of analysis and communication to his university service as he does to his research. Soft spoken but committed, the room quiets to a hush when he speaks as people strain to hear his views.

The University of California condensed matter theoreticians have a tradition of meeting once a year to exchange information in a nice two-day retreat, although the past several years has seen a hiatus in these meetings. Typically, we would meet at the UCLA Conference Center at Lake Arrowhead in the San Bernardino Mountains of southern California for many good hours of talks and dialogue and generally good camaraderie. Walter and his wife Mara would attend many of these meetings and it offered a good opportunity to talk to them on a range of issues ranging from science, to politics, to art and music. They are a wonderful couple and their love and respect for each other is unmistakable.

My wife, Gail, and I have been fortunate to strike up a good personal friendship with Walter and Mara and we've visited them at their lovely home in the hills around Santa Barbara, a home with a beautiful view of the ocean. Many of the readers may not know that Mara Vishniac Kohn is the daughter of the late Roman Vishniac, a world renowned photographer – those interested will find lots of interesting information on Roman Vishniac on various web sites.

The Kohn home is filled with many artifacts of Roman Vishniac's life and works that Mara, with some help from Walter and others, has spent many fruitful hours collecting and archiving. There is a nice museum quality to the Kohn home in Santa Barbara that Walter and Mara graciously shared with Gail and me. The interested reader can find a fairly recent book that Mara edited entitled: *Roman Vishniac – Children of a Vanished World.*

As a closing remark I would like to emphasize how wonderful so many people felt when Walter won the Nobel Prize in Chemistry in 1999. It was good enough for us physicists that this great and noble person won the Nobel Prize, but an added bonus was that his award was presented in an allied discipline, chemistry, so symbolic of the new world of multidisciplinary scholarship proudly symbolized by Walter's achievements and his Nobel award. I'm proud to be his colleague and even more proud to know him as a friend and role model.

**About the Author**: Barry M. Klein is Vice Chancellor for Research and professor in the Department of Physics at the University of California at Davis. His research interests are in condensed matter physics with a focus on computational modeling and interpretation. He has published widely in the research areas of superconducting materials, defects in solids, and in elastic and phononic properties of materials. As Vice Chancellor for Research at UC Davis he oversees the campus research activities which this past year encompassed research awards in excess of $356M. Contact: Department of Physics, University of California, One Shields Avenue, Davis, CA 95616, U.S.A.; bmklein@ucdavis.edu; ovcr.ucdavis.edu/vc/VC.html

# Many Happy Returns, Walter!

Norman Kroll

University of California, San Diego, U.S.A.

I first became aware of Walter Kohn as the coauthor of the Borowitz and Kohn paper on mesonic effects on nucleon electromagnetic properties, one of the pioneering attempts to apply the newly developed quantum field theory methods which had been so successful in quantum electrodynamics (QED) to nucleon-meson problems. That particular problem had attracted quite a bit of attention and had a number of twists which could provide more anecdotes, but in another place. I had just come to Columbia as an Assistant Professor and was franticly preparing for my maiden teaching assignment. I had spent the preceding academic year (1948–1949) as an NRC fellow at the Institute for Advanced Study, a marvelous year in which Freeman Dyson brought Feynman's techniques to the masses and presented a systematic approach to extending renormalization techniques to higher order perturbation theory, and in which everyone else was working on their applications. I had spent the years before that sharing an office with Willis Lamb at the Columbia Radiation Lab where I had the privilege of watching the emergence of the detection and accurate measurement of the fine structure of the Balmer line of atomic hydrogen (the famous Lamb Shift) and collaborating with Willis on an old fashioned (but correct) calculation of its magnitude.

I remained at Columbia through the fifties. It continued to be one of the most exciting places to be, and in many ways I enjoyed living in New York. I had, however, heard of the ambitious plans for a new UC campus located at La Jolla, aware of their success in luring George Feher away from us, and was quite intrigued when Keith Brueckner asked me if I could be interested in a position there. The decision to move was a difficult one for me (although never regreted) and required two years. The final stages of persuasion were carried out by Walter, who had taken over as chair when Keith went on leave, and it was in that context that Walter and I became acquainted. The persuasion stage lasted three months in situ because I had requested that my initial appointment be as visiting professor on leave from Columbia, and provided the opportunity for the growth of our friendship.

He and Lois were very helpful in many ways in facilitating our adjustment, and we share many interests outside of science, especially music and art. The Kohns were quite active in such things as the La Jolla Chamber Orchestra and the La Jolla Museum, and succeeded in involving us as well. They were very active in an ultimately unsuccessful effort to link Rosalyn Tureck (a famous specialist in Bach keyboard music) to our music department and another (quixotic in retrospect) unsuccessful effort to establish a museum to bring the Hirshhorn collection permanently to La Jolla (Washington D.C. was the winner).

While we shared in development of departmental as well us university educational programs and campus growth more generally, we also shared in the turbulence and ferment which charecterized the years of our overlap. Especially noteworthy, because of his initiative and leadership and because it meant so much to him, was the establishment of a distinguished program in Judaic Studies. I very much admired Walter's judgment, tact, and negotiating skills even when I did not quite agree with him. It was a great loss to us when he left us to become the initial director of ITP, but in view of his magnificent contribution to its success, perhaps a gain for science.

In closing I want to express my deep appreciation to Walter for coming to my 80'th birthday celebration last April, for his concern with my continuing physical problems, and for the times he has come to visit me at my home when he is in La Jolla. Happy Birthday Walter. You are a great scientist and great human being.

**About the Author**: Norman Kroll joined UCSD as Professor of Physics in 1962 following a nineteen year period on the faculty of Columbia University and currently holds a post retirement position. His principal research area has been particle physics (quantum electrodynamics, quantum field theory, and particle accelerators). Other areas have included free electron lasers, nonlinear optics, and microwave electronics. He is currently somewhat involved in negative refractive index meta-materials and in particle accelerators. Contact: Department of Physics, University of California at San Diego, 9500 Gilman Drive, La Jolla, CA 92093-0319, U.S.A.; nkroll@san.rr.com

# A Postdoc with Walter

Norton D. Lang

IBM, T. J. Watson Research Center, Yorktown Heights, U.S.A.

In the spring of 1967, to my great delight, I was offered a postdoctoral position to work with Walter Kohn in La Jolla, at the then munificent salary of $10,400 per year. I arrived late one night in October. Driving along some dark canyon road, I began to wonder what on earth I was doing there; it was a rather different place from Cambridge, Massachusetts where I had spent most of the prior ten years. As I soon found out, however, Walter and Harry Suhl had built a strongly interactive community in condensed matter theory that was better than many of the well-known groups in the East, and which drew outstanding people from everywhere.

As I look back over the notes of the discussions I had with Walter, I see now how full of original ideas and insights every one of those discussions was. Each one could have provided the theme and basic approach for a major research project at the forefront of the then current understanding of the field. (The field we worked in together was surface physics; I'm sure the same would have been true for any other area.) Walter was always looking at the most basic questions. He had simultaneously a very deep understanding of both the physics and the mathematics of the problems we discussed; he was always searching for some kind of universal behavior. He focused on major problems and looked for elegant solutions. Mathematical rigor was always important. Even though Walter had, so far as I could tell, never used or programmed a computer, he had a very sophisticated understanding of the numerical analysis which was at the core of the computational work that we did. Collaborating with Walter was an unforgettable experience. I felt that I was working with one of the great physicists of our day (or any day); only much later did I find out that I had really been working with a renowned Viennese chemist.

It is my great pleasure to wish Walter the very best on his 80th birthday, and to wish him as well continued good health and enjoyment of science, to which he has made such profound contributions.

**About the Author**: Norton Lang is a member of the staff at the IBM Thomas J. Watson Research Center. His work has been mostly in surface science, and more recently in the area of molecular electronics. He shared the Davisson-Germer Prize of the American Physical Society for surface physics with Walter Kohn in 1977. Contact: IBM Research Center, Yorktown Heights, NY 10598, U.S.A.; LangN@us.ibm.com

# Reminiscences on the Occasion of Walter Kohn's 80<sup>th</sup> Birthday

James S. Langer

University of California, Santa Barbara, U.S.A.

Walter first emerged in my life in the fall of 1954. I was in my last year as an undergraduate physics major at Carnegie Tech (now Carnegie Mellon University), and Walter was a junior member of the faculty. I no longer remember whether Walter volunteered or was drafted for the responsibility, but somehow he became my private instructor for a year-long supervised reading course in quantum mechanics. We went through the first edition of Schiff's classic text essentially from cover to cover, doing most of the problems and often talking about the ways in which those ideas were being used in current research.

I recall many such discussions, many times when I didn't think I was catching on as quickly as I thought I should, and many other times when I felt that I was being given keys to the secrets of the universe. I especially remember the time when Walter arrived at his office with an intense gleam of triumph in his eye and announced that he had just done something "fiendishly clever." The achievement was possibly his calculation of the energy levels of shallow impurities in semiconductors, a calculation that I could have followed by that time, and possibly did. His enthusiasm made a lasting impression. I never shall forget his joy in the discovery and my realization that I might someday share that kind of experience.

It was lucky for me in many ways to have been adopted by Walter at that early stage of both of our careers. I learned during the course of the year that I had won a Marshall Scholarship, good for study at any university in Great Britain that would accept me. Walter pointed me toward Peierls at Birmingham, where Walter had spent some time a year or so earlier. From then on, our reading course became preparation for me to go directly into thesis research on the English schedule, skipping the year or so of classes that would have been required had I gone to graduate school in the US or done the second undergraduate degree in England that was common for American students.

Peierls' Department of Mathematical Physics at Birmingham was an incredibly wonderful choice for graduate school, and Walter was one of relatively few undergraduate advisors in the US who would have been able to tell me about it at the time. For just a few years, it was the center of theoretical physics in Europe, perhaps more active than even Copenhagen or Cambridge. When I arrived in the fall of 1955, the Birmingham group consisted of Peierls and a few younger staff members (including, for example, Gerry Brown, Dick Dalitz and Sam Edwards), a small number of research students like myself, and an ever changing stream of visitors from all over the world. I shall say more about it later.

In 1958, when I finished my Ph.D. thesis, Walter brought me back to Carnegie Tech under false pretenses. He spent part of that year in London, and I took the train down from Birmingham one day to talk about plans for the future. He explained that he was probably taking a job at Penn the next year, but advised me to accept a position at Carnegie to work on particle field theory with Gian Carlo Wick. (Walter, of course, knew that I had a romantic reason for wanting to return to Pittsburgh.) By the time I got there, Wick had left for Columbia and Walter had decided to stay at Carnegie for another year or so. So I joined the 1958 version of the Kohn solid-state theory group.

Those were the days in which Walter was writing his famous papers with Quin Luttinger. Sy Vosko was at Carnegie working on many-body theory; and John Ward was there for a while, leading us off enthusiastically in highly mathematical directions. We were trying to understand how to solve interacting-electron problems, and most of us were doing this by pushing diagrammatic techniques beyond their limits of validity. Walter was much interested in this project and contributed to it in essential ways, but I do not think he was ever so enamored by those formalisms as were many of the rest of us.

Walter then moved to La Jolla and, a year or so later, I went there on what amounted to a delayed postdoctoral fellowship. Walter's basic style was obvious by that time. He always has loved mathematical elegance, but he reserves it for situations where it is truly necessary. His emphasis then, as now, was on the most important physical questions and the ways in which they could be answered with insight and confidence. That was 1961–62, and he was moving toward density-functional theory.

After my year in La Jolla, Walter and I were only occasionally in touch with each other until 1979, when when he agreed to become the founding director of the Institute for Theoretical Physics in Santa Barbara. It was then that he started talking with me about the possibility of joining him there as one of the permanent members of the institute. I remember a discussion in a restaurant one evening during some physics conference. In typical Walter fashion, he had thought deeply and carefully about the role that such an institute might play in advancing modern science. He also had thought just as deeply and carefully about every detail of how it would be organized and operated. I became keenly interested. The clincher came in February 1981, when Walter arranged that my wife Elly and I would visit sunny Santa Barbara and go hiking with him at a time when the weather in Pittsburgh was most cold and dreary.

We moved to Santa Barbara in the spring of 1982. Since then, Walter and I usually have occupied offices just a few doors down the hall from each other, but we have worked together primarily in institution-building modes. Our discussions more often have been about human rights or issues of science and society than they have been about condensed matter or materials physics. (We always have been interested in related research areas, but somehow we never have co-authored a scientific paper.) Our main project, of course, was the ITP, which is such a big story – with Walter as the essential central figure – that it needs more careful telling than seems appropriate here. Let me try to convey just a bit of the flavor of that enterprise by connecting it to Walter's earliest influence on my career.

Bob Schrieffer moved to the ITP the year before I did. He and I had met in 1957 in Birmingham, where he had come for postdoctoral study after finishing his famous Ph.D. thesis. Paul Martin was the first chair of the ITP Advisory Board; he played an active role in establishing the institute and recruiting its first permanent members. Paul also had been a postdoc in Birmingham while I was a graduate student there, and we had come to know each other quite well. Thus all four of us – Walter, Bob, Paul, and I – had seen first-hand what Peierls had accomplished; and all of us explicitly hoped to create something similar in Santa Barbara. Along with the remarkable "gang of four" (Jim Hartle, Ray Sawyer, Doug Scalapino, and Bob Sugar) who really got the institute started at UCSB, we wanted to build a truly international center for research, with free exchange of ideas across all the physics-related disciplines, and easy, supportive interactions between established scientists and beginners.

I have been too close to the center of this enterprise to be an objective judge of its success. And I realize all too well that it has been almost half a century since Walter sent me to Birmingham, and that the world of physics is far larger and much different now than it was then. So it is hard to see from the inside how Peierls' department in Birmingham and Walter's ITP have truly influenced the course of scientific history. In another fifty years, Walter certainly will remain known for his Nobel Prize winning density functional theory. I expect, however, that he will be equally well remembered for his leadership of the ITP, for his guiding influence on young scientists, and for the standards he has set in the conduct of modern research.

**About the Author**: James S. Langer is professor of physics at the University of California, Santa Barbara, where he was director of the Institute for Theoretical Physics from 1989 though 1995. He was president of the American Physical Society in 2000, and is currently vice president of the National Academy of Sciences. His research interests have been in theories of nonequilibrium phenomena such as the kinetics of phase transitions, pattern formation in fluid dynamics and crystal growth, earthquakes, and most recently the dynamics of deformation and fracture in solids. Contact: Department of Physics, Broida Hall, University of California, Santa Barbara, CA 93106, U.S.A.; langer@physics.ucsb.edu

# Some Recollections About Walter and His Work

David Langreth

Rutgers University, Piscataway, U.S.A.

My first encounter with Walter's work was as a graduate student at University of Illinois in the early 1960's, while taking David Pines's course in advanced solid state physics. There we learned about the effects of the conduction electrons in a metal on the phonon spectrum, and in particular about the features in the spectrum that can be caused by the sharpness of the Fermi surface[1] that were dubbed "Kohn anomalies" by their experimental discoverers[2] a couple years later. We also learned about Walter's work with Sy Vosko on the effect of impurity induced oscillations in the electron density on the Knight shift in nuclear magnetic resonance.[3] All this came back into focus a couple of years ago when I introduced Walter at a joint physics-chemistry colloquium at Rutgers, where I mentioned that Walter's work had already been available at the textbook level during my graduate school days around forty years earlier. Walter immediately expressed skepticism that these contributions could have reached textbooks so fast, so I later felt bound to check on my memory. Yes, Walter, it really had happened. My course textbook[4] in David Pines's course, published in the year of your fortieth birthday, discusses both of the above topics. In addition, Charlie Slichter's book,[5] which was used as the textbook for the magnetic resonance course taught at Illinois at the time, discusses your work in that area.

My next encounter with Walter's work was during the 1964/65, a postdoctoral year for me at the University of Chicago. There were a large number of postdocs and graduate students in solid state theory at Chicago that year, and we had journal clubs and active discussions of the interesting papers that appeared, a much easier task in those days, when the number of papers per year was much smaller than it is today. But it was a banner period for the ap-

[1] W. Kohn, Phys. Rev. Lett. **2**, 393 (1959)

[2] B.N. Brockhouse, K.R. Rao, and A.D.B. Woods, Phys. Rev. Lett. **7**, 93 (1961)

[3] W. Kohn and S.H. Vosko, Phys. Rev. **119**, 912 (1960)

[4] D. Pines, *Elementary Excitations in Solids*, (Benjamin, N.Y., 1963)

[5] C.P. Slichter, *Principles of Magnetic Resonance* (Harper & Rowe, N.Y., 1963)

pearance of papers that have had timeless influence. In particular, there was the famous Kondo paper[6] explaining the long mysterious electrical resistance minimum in certain metals, and suggesting a new weakly bound ground state involving spin correlations, that has so far provided nearly forty years of research interest fundamental to understanding condensed matter systems with strong on site correlations. There were also the famous Hubbard model papers[7] which formulated the fundamental dilemma in the proper understanding of such highly correlated systems, along with the beginnings of methods to treat them. I can remember extensive discussions of the Kondo and Hubbard papers, and the realization by all of us that these papers were destined to have a long standing impact on condensed matter physics. We also found and discussed a paper by Hohenberg and Kohn[8] which proved the existence of a universal electronic density functional and the variational bound it satisfied. We found this paper fascinating, but none of us expected this work to have a long-lived impact, and I certainly didn't dream that I might be working in this area some ten years later, and then off and on for the balance of my professional life to date. Walter told me recently that he too did not at that time anticipate the impact either. When asked when he did begin to anticipate the impact, he said that it was with the paper with Lu Sham[9] and afterwards. This is the work that derived the self-consistent equations, now known as the Kohn-Sham equations, which provide the practical method by which density functional theory is still normally implemented to this day. I conclude this section by mentioning another paper[10] which, although written earlier, was unknown to me before this postdoctoral year. It contributed to our understanding of the interaction of electromagnetic fields with condensed matter, and I have looked back at it a number of times over many years.

Despite my familiarity with some of Walter's work, I didn't meet him in person until the next decade. I don't remember for sure precisely when we first met, but it may have been at a memorable conference, a Nobel Symposium in the suburbs of Göteborg, Sweden, June 1973. By that time the paper with Norton Lang[11] applying density functional theory (DFT) to planar metallic surfaces had been long since in print and the success of DFT was well on its way to being established. I can remember the fine reception that Walter's presentation received. A number of other spectacular advances were presented at this conference. On the theory side was the presentation of the successes of renormalization group theory both in critical phenomena and in the numerical solution of the Kondo model. On the experimental side was the presentation of the discovery of superfluid liquid $He^3$. The latter two discoveries went on

---

[6] J. Kondo, Prog. Theor. Phys. **32**, 37 (1964)

[7] J. Hubbard, Proc. Roy. Soc. A **276**, 238 (1963); **277**, 327 (1964); **281**, 401 (1964); **285**, 542 (1965)

[8] P. Hohenberg and W. Kohn, Phys. Rev. **136**, B 864 (1964)

[9] W. Kohn and L.J. Sham, Phys. Rev. **140**, A1133 (1965)

[10] V. Ambegaokar and W. Kohn, Phys. Rev. **117**, 423 (1960)

[11] N.D. Lang and W. Kohn, Phys. Rev. B **1**, 4555 (1970)

to win Nobel prizes in physics. As DFT went on to success after success in predicting the ground state properties of large classes of solids and solid surfaces, it was expected by many of us that a Nobel prize for DFT would follow. But it was destined to wait for the many improvements in the accuracy of the functional on the finer energy scales required to make it useful for chemical applications.

The academic year 1973/74 I spent my sabbatical at La Jolla (UCSD). My visit was arranged through Lu Sham and Harry Suhl and I don't remember interacting with Walter in any strong way that year. However, I especially remember the wonderful brown bag lunches which Walter organized weekly. The format varied from discussion of the research of the people there, to interesting physics in preprints or journals, and even to current events happenings, and were attended by a rather wide variety of physics people. I remember especially interacting with some of the outstanding young people, graduate students and postdocs, many of whom went on to become outstanding research physicists, including John Dobson, Werner Hanke, and Eugene Zaremba. My recollection is that of these, only Eugene was formally associated with Walter at that time; but what a time it was, when the theoretical understanding of the van der Waals interaction between an atom and a surface was in the works.[12] This was work that affected my own, but not until two and a half decades later.

My period of strongest interaction with Walter came in 1983, as Walter had asked me, along with Dennis Newns and Harry Suhl, to coordinate an eight month program on many-body phenomena at surfaces at the Institute for Theoretical Physics (ITP), in Santa Barbara. This was the year of Walter's sixtieth birthday, and his good friends had arranged a wonderful party for him, which I will never forget. Sometime early in the evening a guest whose name I don't remember engaged me in conversation with the following questions,

– Were you a student of Walter's?

– No.

– A postdoc?

– No.

– A collaborator?

– No.

– A close friend?

– No.

– What are you doing here, then?

It did not seem to satisfy him when I explained that we were planning and coordinating a program with Walter, or even that all ITP members had probably been invited. He seemed ready to escort Ellen and me to the door as freeloaders of food and drink. This story came to my mind instantly, when I was asked to write this piece, because the answers to the questions have not

---

[12] E. Zaremba and W. Kohn, Phys. Rev. B **13**, 2270 (1976)

David Langreth

Walter at Rutgers in May 2001 shortly before the degree presentation ceremony

changed in the intervening two decades. Friends, of course. Close friends, I am not so lucky.

Now, back to the ITP program in many-body phenomena at surfaces. Even in the planning stages the year before, I began to feel that it was a program dear to Walter's heart. However, I learned soon after arriving in Santa Barbara that Walter treated all programs with tender loving care. As far as I could tell, Walter went to every talk, and used his great breadth to really participate in all of them. As far as I could tell every member would be invited to Walter's office for private discussions of his or her research. Every member was invited to lunch with Walter at least once. Every member was invited to his home. I feel that we really owe a great debt of gratitude to Walter, who as the founding director of ITP gave five years of his life to make ITP a resounding success.

While I mark in my mind Walter's sixtieth birthday with the above ITP program, I mark his seventieth birthday (a year late) with a shorter ITP program during the summer of 1994. The topic was specifically density functional theory, and it marked in my mind a fundamental change that had taken place during the previous ten years. Although there had for a long time been chemists active in the DFT field, it was clear both by the attendance and by the topics of active interest, that the center had now shifted toward adapting DFT to problems in chemistry, and that the functionals had become sufficiently accurate to make this a profitable endeavor. And I could see Walter actively participating with the chemists in the intellectual development.

Walter's seventy-fifth birthday year was marked by the Nobel announcement that we are all so happy about. I started to think that it would be nice if Rutgers were to offer an honorary degree to Walter. Of course Walter had won many honors and received many honorary degrees. After a moment's thought, I realized that one more could hardly mean very much to him.

Nevertheless, I went ahead and contacted Walter asking him whether he would be willing to honor us in this way. I was delighted when he said that he would. Then, with the help of many colleagues in chemistry and physics, including especially Kieron Burke, we succeeded in arranging the invitation, although it was to come later than we originally planned. And so Walter spent a number of pleasurable days with us in May 2001 (see Fig. 1).

And so I conclude by wishing you, Walter, the very best on your eightieth birthday. For me, it also marks the fortieth anniversary of when your work began to have an impact on me.

**he Author**: David Langreth is a professor of physics in the Center for Materials Theory and Department of Physics and Astronomy at Rutgers University. His research spans a wide range of topics in condensed matter theory, surface physics, materials physics, and nanoscale physics. His current interests include the development of density functionals which account for long range van der Waals interactions, and the dynamical properties of quantum dot single electron transistors in the Kondo regime. Contact: Physics and Astronomy, 136 Frelinghuysen Road, Piscataway, NJ 08854-8019, U.S.A.; langreth@physics.rutgers.edu; www.physics.rutgers.edu/~langreth/

# Romancing the Theorem

Mel Levy

Tulane University, New Orleans, U.S.A.

It gives me joy to recall here a few of my interactions with Walter and his wonderful theorem, especially since perhaps I have spent a larger fraction of my time than most in analyzing this gem in a formal sense. It has been a love affair.

Leaps in science require, of course, that the right questions be asked. In view of this, it astounds me that no one prior to 1964 inquired, "What is the *definition* of the exact functional that Thomas and Fermi tried to approximate". We are all fortunate indeed that Kohn asked this simple but profound question. The Hohenberg-Kohn Theorem that followed has allowed us to feel much intellectual enjoyment, has led to exciting practical applications in a number of fields, and has literally enabled so many of us to pay our bills by doing what we love! Had there been no Kohn to ask the question in 1964, there is no guarantee that it would have even been asked by now. Thanks, Walter.

1965/66 was my first academic year of graduate study in chemical physics at Indiana University. That was when I might have seen Walter for the first time. I very briefly attended a conference at the University on many-body theory. I believe that the only lecture I heard at this meeting was Uhlenbeck's. For personal reasons, I had to leave right after his talk. I thus missed the lectures of Kohn, Bethe, and others that followed. Thus, my first interaction with Professor Kohn consisted of walking out before his lecture on his new density-functional theory. Fortunately, things got better after that.

It was in 1972, at the beginning of my postdoctoral studies with Bob Parr at Johns Hopkins, that I learned from him about the existence of some theorem that connected the energy with the electron density, the Hohenberg–Kohn Theorem, of course. However, I did not study its proof in depth until the late 1970's, as a new faculty member at Tulane University.

The next time I saw Walter was in the summer of 1975, at a theoretical chemistry conference in Boulder, Colorado. A special session was spontaneously arranged concerning the HK theorem. During this session, the validity of the theorem was challenged. I don't remember what the issue was. I do remember, however, that Walter responded by ending the session with a proof of the theorem on the blackboard. An ovation followed. Nevertheless, some skepticism existed concerning the usefulness of the theorem for quantum chemistry. One professor told me that he thought that the emerging interest in DFT was just a "passing fad". Also, quantum chemists were not accustomed to $N$ orbitals for $N$ electrons, as employed in Kohn–Sham theory. It was felt that something was missing if just $N$ orbitals were employed. Of course, $N$ orbitals are just what we want most of the time.

In 1976, I came to Tulane. A year later John Perdew joined us. He presented a special topics course on DFT and I sat in on some of the lectures. As a result of this course, I started to analyze Kohn–Sham theory for degeneracies, which led to the 1982 Phys. Rev. A publication on fractional occupation numbers and non-$v$-representability.

In December of 1978, I studied the proof of the Hohenberg–Kohn theorem in earnest. Between semesters, my wife Michele took our two year-old son and three month-old daughter to visit our parents in Washington and New York. This was very tough on her. The object was for me to be left alone in order to write a grant proposal, a desirable deed, especially for an untenured faculty member. I never wrote that proposal, which did not please Michele very much at the time. Instead, I utilized that opportunity to stare at and ponder the HK theorem. This led to the 1979 PNAS publication of the constrained-search proof of the generalized HK theorem, which extended the original HK functional to densities that arise from certain degeneracies. I was certainly very excited about this work. But what is most important here is Walter's interest in it. It thrilled me, as a young faculty member, to have such strong support from one whom I admired so much. It is often the case that someone of stature would simply ignore at best a new approach by a junior researcher. I am grateful that his support and encouragement has lasted all these years in different forms for different purposes. Of course, I have plenty of company with those feelings.

I first spoke with Walter at length in Santa Barbara in 1983. I also interacted with Lu Sham for the first time there. Walter and I have had countless interactions since. It is difficult for me to reveal anything about his creativity and thinking process that is not already well known. He has inspired us to think and write better through his scientific articles and through his tireless and relentless critical analysis of each of our lectures at meetings. No faulty argument or shred of sloppy reasoning gets by. And this has been true lecture after lecture, year after year, to this very day.

A description of a memorable conference in Crete in the summer of 1998 is given by Michele in the account that follows this one. I last saw Walter at a workshop in Albuquerque late last summer. We reminisced about earlier days in DFT during an afternoon after the workshop, while he was waiting for his ride to Santa Fe. He looked especially well and happy. Wishing you at least twenty more years of this, Walter!

Mel Levy

**About the Author**: Since 1976, Mel Levy has been a professor in the Chemistry Department and Quantum Theory Group at Tulane University, near the bayous and the swamps. He is presently on leave as Research Professor of Chemistry and Physics at North Carolina A & T State University, between the Blue Ridge Mountains and the Atlantic Ocean. Since the early work described above, he has enjoyed studying the exact properties of the functional and its derivative for approximation purposes. Contact: Chemistry Department, Tulane University, New Orleans, LA 70118, U.S.A.; mlevy@tulane.edu

# Crete, June 1998

Michele Levy

North Carolina A & T State University, Greensboro, U.S.A.

During the International Workshop on Electron Correlations and Materials Properties, June 28–July 3, 1998, near Iraklion on Crete (the last meeting Mel attended before Walter received his Prize, and the only one whose beautiful poster we framed), I hung out with the spouses and friends. In our ample free time, we explored the capitol city or joined organized tours that left from Amoudara, our tourist beach outside Iraklion. We made a pilgrimage to the tomb of Kazantsakis, banished outside the city walls, and visited the Lassithi Plain (home to the Valley of the Windmills and the Dictean Cave, birthplace of Zeus), Gortys (the old Roman capitol of Crete), the caves at Matala, and the ruins of Phaistos on the Mesara Plain.

Having all week long endured my praise of Crete's beauty, Mel wanted to see some of the island, too. So I arranged a rental car for after the conference. By Friday, many of the attendees were leaving. But Walter and Agnes Nagy had stayed on. So the four of us decided to travel somewhere together. I suggested that we head south, toward Matala. This way we could also take in Gortys and Phaistos.

It was another day of piercing sun – a heat wave was setting records all over southern Europe. Walter looked particularly jaunty in his cap and seemed ready for anything. So off we went, with hats for shade, water for thirst, in the spirit of adventure. It might have reached 113 in the sun, but we managed just fine. We lunched on the stone patio of the little shop at Phaistos, under a welcome canopy of trees. Eating our Greek salads, good yogurt, and sandwiches, we had a marvelous view of the mountains and the ruins of the old Minoan palace. Then we wandered about the splendid but treeless and sun-drenched site.

Despite the often-treacherous mountain roads and our lack of familiarity with the map, we made it to the beach at Matala, its hot sands bordering the Libyan Sea. First we visited the caves, once burial grounds for early Christians, later rustic home to sixties hippies. Then we waded into the sea, its cool waters refreshing in the intense afternoon heat. The sand scalded the soles of our bare feet.

Walter with Mel and Michele at Phaistos (Crete, 1998)

By the time we started back, it was near dusk. We worried just a bit about negotiating hairpin-turns on one-lane each direction mountain roads with little light. As we drove, I mentioned to Walter the novel I was writing about the Balkans in 1968, with its East German anti-hero. At his request, I related the tale of my border crossing from Bulgaria into Turkey with this character's real-life counterpart in the Mercedes he was smuggling. Walter's amusement and lively interest spurred me back to work with gusto!

Suddenly from a dirt road strung along the cliff, a battered pick-up truck piled high with watermelons entered our lane just ahead. It lurched along slowly, a function of its age and load. Given the road's narrow width, its constant and dangerous curves, and the absence of anything one could call a shoulder, we dared not pass the truck and settled in behind it, presumably for the duration.

At various points the driver seemed to check his rear-view mirror, and he did keep to the far right of the lane. But to move ahead would be to take a chance on what might come around the corner towards us on the left. We'd all heard stories about horrific accidents on roads like this.

Farther up, the lane widened slightly. The truck pulled over to the right and motioned us forward. Even then we did not feel fully secure; here, too, the road zigzagged sharply around a mountain bend. But we seized the opportunity and reached the hotel safely before nightfall. We duly noted the Cretan's role in our "survival". His thoughtful act seems particularly meaningful, given the events of the following October.

Back at the hotel, feasting on juicy watermelon, I made a mental note of that episode, and later wrote a story, "Still Life and Watermelon," that dealt in part with our day trip and "the watermelon man." Once this work was published in a literary journal, and a Cretan friend told me that I had truly captured her island, I thought how glad I was to have memorialized our weekend excursion together, even in this oblique manner. For it has lived in my memory as a celebration of natural beauty and friendship.

**About the Author**: Michele Levy chairs the English Department at North Carolina A & T State University. Her current interests include 19th century Russian and contemporary postcolonial and Balkan literature. She has also published short stories and, following a visit to Serbia this fall, has a novel being translated into Serbian. Contact: North Carolina A & T State University, English Department, 1601 East Market Street, Greensboro, NC 27411, U.S.A.; mflevy@ncat.edu

# Recollections of Walter Kohn

Steven G. Louie

University of California, Berkeley, U.S.A.

Although Walter Kohn and I never collaborated on a particular project together, his many groundbreaking papers in condensed matter physics have influenced my own career greatly. As is the case with many others, I first heard of Walter when I was in graduate school. In standard courses, we learned about the Kohn anomalies, the Luttinger–Kohn effective mass theory, the Korringa–Kohn–Rostoker (KKR) method, the density functional formalism, and on and on. I was in awe of the achievements of this great physicist. It was not until I finally met Walter in one of the APS March Meetings in the mid-1970's that I realized how approachable and friendly a person he is. I don't recall exactly which March Meeting it was, but I can still vividly remember my first encounter with the hallmark "giggly" laugh alluded to in the Preface of this Volume and the many insightful, deliberately delivered comments and questions he offered.

Walter's papers are always a joy for me to read. They are full of physical insights and written in such a manner that they are comprehensible to any serious theory students. Nowadays, whenever I teach density functional theory (DFT) to my graduate quantum theory of solids class, I strongly encourage the students to read the two original classic papers by Hohenberg and Kohn and by Kohn and Sham on the subject, especially the latter paper which provides a practical, accurate framework for treating electron-electron interactions in modern electronic structure calculations of the ground state.

In the 1970's and 1980's, we were working at Berkeley on methods that would be predictive for structural energies, phase transitions, and vibrational properties. The combination of the *ab initio* pseudopotential approach and DFT turned out to be a real winner. It ushered in an era in which researchers could accurately compute and sometimes even predict the structure of crystals, surfaces and defects, and a range of other properties of solids from first principles. The usefulness of results from DFT calculations of course extends far beyond the realm of the study of structure and energetics. For example, the ability to compute the full dielectric matrix greatly benefited my own work on studying excited-state properties within the GW approach. It was one of the key ingredients that allowed Hybertsen and me to develop an *ab initio* method to calculate the band gaps and quasiparticle energies in semi-conductors in 1985.

I got to know Walter better personally over the years, especially through some of the all UC condensed matter theory meetings at Lake Arrowhead in the San Bernardino Mountains in Southern California where theorists from all the campuses of the University of California would meet for a relaxed weekend of physics and socializing. Our most recent extended encounter was in the summer of 1998 at a conference on electron correlations and materials properties in Crete. The participants were basically trapped in a 5-star resort with all meals served. Every morning Walter would join us at the breakfast tables with a broad smile and sun hat in hand. In this setting, we not only discussed physics, we had a chance to chat with Walter on an astonishing range of subjects that he is interested in, from history to travel to the arts. It was great fun.

Walter is always a gentleman, kind and inspiring to the younger people. I remember, in the fall of 1998, I tried to invite Walter to give the opening talk at the 9$^{th}$ International Workshop on Computational Condensed Matter Physics in Trieste scheduled for January of 1999. Amidst all the excitements of his Nobel Prize and concerns with the illness of Mara, Walter still diligently answered all my emails. Although finally he was unable to come, I really appreciated his efforts and kindness during that most busy time of his life.

Walter Kohn is a towering figure in theoretical physics. As a scientist and human being, he has inspired generations of young theorists. I am very fortunate to have gotten to know Walter personally.

**About the Author**: Steven G. Louie is Professor of Physics at the University of California at Berkeley and Senior Faculty Scientist at the Lawrence Berkeley National Laboratory. His current research interests are in theoretical condensed matter physics and nanoscience, covering the areas of electronic and structural properties of solids and nanostructures, quasiparticle and optical excitations, correlation effects in bulk and reduced-dimensional systems, and superconductivity. He is the recipient of the Rahman Prize for Computational Physics and the Davisson-Germer Prize in Surface Physics of the American Physical Society. Contact: Department of Physics, University of California, Berkeley, CA 94720, U.S.A.; sglouie@uclink.berkeley.edu

# Walter Kohn – Points of Impact

Bengt Lundqvist

Chalmers University of Technology, Göteborg, Sweden

Today also persons outside your immediate neighborhood can influence your life. To me Walter Kohn is such a person. Our paths have crossed just about a dozen of times, including conference encounters, but the impact is certainly there. I'll try to describe that briefly by a few points of impact.

*1961–69*: As a graduate student of solid-state physics in the early sixties (the term condensed-matter physics was hardly invented then), I of course came across "Kohn" at many occasions – the Kohn anomaly, Korringa–Kohn–Rostoker method, ... – in discussions and in course textbooks by Kittel, Ziman, Nozières, ... . In my first Physical Review paper, the damping of phonons in aluminum was calculated, with effects closely related to the Kohn anomaly. Otherwise the single-particle spectrum of the interacting electron gas kept me busy.

*1967–69*: However, close to the end of my graduate studies my supervisors, Stig Lundqvist and Lars Hedin (deceased in 2000 and 2002, respectively), brought home the Kohn–Sham density-functional theory (DFT) to Chalmers. Having an interesting structure in the single-particle spectrum of the interacting electron gas, I was initially not too enthusiastic about deriving "interpolation formulas" for the exchange-correlation (XC) energy and potential, but fortunately I finally "agreed" to do the 'Hedin–Lundqvist density functional'.

*1971*: During a visiting-scientist year at Cornell I paid a visit to Lu Sham and Walter at La Jolla. In addition to great physics discussions, seminar on the "Swedish electron gas" (John Wilkins' terminology), great hospitality of my host, and beach experiences, I particularly remember the Kohn-group Friday-lunch seminar. It was very educational to experience Walter's working mode. With about a dozen graduate students and post-docs meeting up, one or two of them presented an on-going project, and then we all got a two-hour performance from Walter. No aspect of the problem under consideration was unimportant enough to avoid his critical analysis and constructive contributions.

*1973*: We arranged a Nobel symposium on "Collective Properties of Physical Systems" at Aspenäsgården, in Lerum outside Göteborg. 'All' the big international stars in the field were gathered and given the "full treat" (including folk-dance and banquet (comment: At just about the time when I said "they've laid on just about everything but dancing girls," in came the dancing girls.). Walter presented "Theory of Surfaces of Simple Metals", the now canonized jellium results and figures with Norton Lang (and I got a chance to mention the role of spin polarization for the first time). In his conference summary, P.W. Anderson mentioned the "monumental work that Walter Kohn described", which he commented on further: "In surfaces there was the monumental paper and sequence of work that Walter Kohn has assembled on this subject. I personally had not realized how close one could get agreement with experiment. It seems the basic equilibrium processes that happen on surfaces are beginning to be very straightforwardly understood. There do of course remain problems. I think the discussion session will be valuable on the question of the plasmon contribution." The statement was certainly right at the time (accounting for, e.g., trends in workfunction values was impressive and is still significant), but it can also be given a visionary interpretation. I think it can be expressed in almost the same way today, however, on a radically different quantitative basis. After Walter's talk some of us composed the song "Look how good the numbers are" (I have (un)fortunately forgotten the melody (and text)), reflecting the discussion of why the local-density approximation (LDA) is applicable at a strong inhomogeneity like a surface.

A small fraction of the participants of the 1973 Nobel symposium on "Collective Properties of Physical Systems" at Aspenäsgården, in Lerum outside Göteborg. Walter Kohn can be found in the front row, together with Harry Suhl and David Langreth. Other identified participants are Bengt Lundqvist, Bob Schrieffer, David Mermin, Seb Doniach, P.W. Anderson, I. Dzyaloshinsky, Stig Lundqvist, Ivar Waller, L.P. Gorkov, Armand Lucas, Norman March, Vinay Ambegaokar, Claes Blomberg, and John Hopfield

At this visit I learned that Walter had old friends in Sweden, including a close friend in high school in Vienna, which we felt increased our chances to get the great man to Sweden more times. Later I have learned that caring about old friends is a characteristic virtue of Walter's. I also learned about his broad interests. He seemed to like the 'social program' (opera, ballet, singing, folk-dance, excursions, the Smetana room of the Göteborg city hall), ...) of our symposium just as intensely as the scientific sessions and discussions, which means quite a lot.

*1983*: An Institute for Theoretical Physics (ITP) has been formed at UCSB in Santa Barbara, and Walter had become the Director of ITP. In the array of impressive ITP activities there was an eight-month program on "Many-Body Phenomena at Surfaces". It culminated in a two-week workshop, which benefited from Walter's presence, and which in turn resulted in a volume with the same name, edited by David Langreth and Harry Suhl. I was one of the fortunate participants of this ITP workshop, happy to experience both the progress made during the workshop and the efficient ITP leadership of the Director.

*1995*: A symposium on Contemporary Concepts in Condensed Matter Physics was organized at Hjortviken, in Hindås outside Göteborg, in conjunction with Stig Lundqvist's 70th birthday, with a 'pre-school' on Materials Theory and a 'post-school' in Surface Science, both with contributions from prominent scientists. We were fortunate to get Walter in the long list of symposium participants that had chosen to show up. And guess who took a very active part in the "post-school" in Surface Science, present at all talks and very active in the discussions? The participating graduate students were more than happy for Walter being there!

Afterwards I drove Walter to the railway station, and we got some opportunity to talk about life and similar things. Stig was a former jazz trumpeter and piano player, and as a tribute to him, the symposium dinner had a surprise concert by a jazz big band, where I am fortunate to play the piano. Waiting for the train, Walter expressed his appreciation for physicists that do more than just physics, a statement that of course made me happy.

*1998*: This was the year, when the decision on the Nobel prize was taken. Stig and I were among the many to express our warmest congratulations for Walter's long-deserved Nobel prize, asking him to stop by in Göteborg before Christmas! We were fortunate to get the kindest of answers: "It's always great to hear from Sweden, whether it is the Swedish Academy or good friends!" and "They are very greatly appreciated. Without the great Swedish interest and contributions DFT might still not be broadly accepted. Best wishes to both of you." Needless to say, such generous sharing gives warmth!

*1999*: With some delay, in June 1999 we could arrange the Kohn Symposium on "Manifestation of applications of Density-Functional Theory in Physics and Chemistry in Northern Europe" in honor of the great Nobel laureate. At this occasion 'Northern Europe' included not only the Nordic countries but also Berlin (Matthias Scheffler), Stuttgart (Ole Krogh Andersen, Michele Parrinello), and New Jersey (Kieron Burke). Walter talked about "Electronic Structure of Matter: Wave Functions and Density Functionals", essentially his Nobel lecture. One nice feature was that chemists and physicists met up in great number and that their coming together was well catalyzed by Walter and DFT. His only personal request, that the widow of his close friend in high school in Vienna should be included in some appropriate social occasion, was taken care of.

*2000*: When Stig passed away, this caring side of Walter's was shown again: "Please convey to Stig's family and Swedish friends my sincere condolences on his death. His deep commitment to scientific progress, especially in the Third World, his encouragement of young persons, and his personal warmth and fortitude will remain a great inspiration to those of us fortunate to have been among his friends."

*In summary*: Like in DFT, there is both exchange and correlation between Walter's and my scientific and personal lives. I am certainly very happy about this fact (and I know that I can speak for Stig and Lars and say that the same applied to them). As indicated in this short sketch, the interactions have been mostly nonlocal. We both know very well that such dependencies are of key importance, not only for van der Waals interactions.

However, local effects are also important, and I hope that my small points of impact above help to illustrate this. I am fortunate to have met Walter, to have been influenced by him, and to be able to congratulate him on his 80th birthday in this way.

**About the Author**: Bengt Lundqvist is professor of Mathematical Physics at Göteborg University (GU), heading the Materials and Surface Theory group and the Department of Applied Physics, Chalmers/GU. His research concerns theory of condensed matter, materials, and surface phenomena. His current interests include developing density-functional theory to account for van der Waals forces, soft matter, non-adiabatic effects in molecule-surface processes, growth and catalysis, nano-scale electronics, and bridging time and length scales. Contact: Department of Applied Physics, Chalmers University of Technology, S-412 96 Goteborg, Sweden; lundqvist@fy.chalmers.se; fy.chalmers.se/ap/mst/

# KKR – Reminiscences About Walter Kohn

Gerald Mahan

Penn State University, University Park, U.S.A.

Many years ago I was a professor at Indiana University. In 1980, plus or minus a few years, I and many others attended the annual meeting of the Midwest Solid State Theorists. That year it was held at Ohio State University. The theorist at OSU put on a great meeting. It was a two day event, with a banquet on the night in between. The banquet was an occasion to honor Prof. Jan Korringa, who was retiring after an illustrious career at OSU.

Korringa is remembered as the first "K" in KKR, which stands for Korringa–Kohn–Rostoker. It is a theoretical framework for doing band structure calculations, and many of the earliest electronic structure computer codes used KKR. Korringa had actually first solved the problem of an electromagnetic wave traveling through a periodic arrangement of conducting spheres. However, the mathematics of that problem is identical to Schrödingers equation for an electron wave traveling in a periodic array of spherical ions. The latter problem was solved later by Kohn and Rostoker.

Walter Kohn was invited to be the after dinner speaker. Walter was in good form, and gave a great speech. He said laudatory remarks about Korringa, and thanked the organizers for putting on a great meeting. He made comments about the state of Density Functional Theory, and other theoretical issues of that time.

Then his speech seemed to turn serious. He said that at one point in his life, he was unmarried and rather lonely. He related that, after a few years, he met a lady and they started to see each other. He became very fond of this lady, and began to wonder if he should get married again. Such an decisison is very weighty, and takes serious consideration. Suddenly he realized that someone, perhaps God, had sent him a positive signal that this was the right lady. The first three letters on her car license plate were "KKR".

**About the Author**: Gerald D. Mahan is a Distinguished Professor of Physics at Pennsylvania State University. Earlier he had faculty appointments at the University of Oregon, Indiana University and the University of Tennessee. He also had research appointments at the General Electric Research Laboratory, and at Oak Ridge National Laboratory. He has published 240 research articles in refereed journals, and several books including Many-Particle Physics. He is a Fellow of the American Physical Society, a member of the Material Research Society, and a member of the U.S. National Academy of Science. Contact: Department of Physics and Materials Research Institute, 104 Davey Laboratory, Pennsylvania State University, University Park, PA 16802, U.S.A.; gmahan@psu.edu

# Cecam's Visitor!

Michel Mareschal

CECAM, Lyon, France

Cecam's connection with Walter Kohn goes back to the seventies. Cecam had just started (it was founded in 69 by Carl Moser) and that was the time when Carl would welcome visitors from the US in order to make computational methods more popular across Europe. Cecam was located inside an important computing center in Orsay and visitors would stay for a few weeks to a few months.

Walter was one of them, visiting Cecam at several occasions during the early seventies and working, as his research reports testify, on the construction of Wannier functions for metallic hydrogen, together with Charles Sommers and Wanda Andreoni. He also participated in discussion meetings and workshops, where discussions were mainly devoted to the comparison of the various local potential based methods, among them the local density approximation to DFT within the Kohn-Sham formalism. People, both chemists and physicists, were starting to realize that those techniques were much more than just an alternative to semi-empirical methods and those most popular among the quantum chemists.

It is sometimes said that Walter took advantage of his stays at Cecam as an experience when he was in charge of starting the ITP. I asked him whether this was true when he visited Cecam, now located in Lyon, for a workshop. He smiled and kindly answered that he had always been very appreciative for Cecam's model and had it present. Probably the influence was minimal!

A true and genuine compliment to Walter's skills came very spontaneously from Cecam's secretary, Emmanuelle Crespeau. She told me after his last visit to Cecam, how nice talking with Walter could be : he can spend an hour having a lively and truly interesting conversation, not only in French but also with no reference to physics whatsoever! «Un grand monsieur !».

**About the Author**: Michel Mareschal is director of CECAM, European Center for Atomic and Molecular Computations, and also a professor at Université Libre de Bruxelles. His research has mainly been in the field of statistical mechanics, at equilibrium and out of equilibrium. His current interests are in the development of numerical methods to describe complex fluid behavior, in particular hydrodynamics at the atomic scale. Contact: CECAM, ENS-Lyon, 46, allée d'Italie, F-69364 Lyon cedex 07, France; Michel.Mareschal@cecam.fr; www.cecam.fr

# Nobel Mania

Ann E. Mattsson

Sandia National Laboratories, Albuquerque, U.S.A.

The night of October 12–13, 1998, was a little different than the same night
the year before. Thomas, my husband, came back to bed already at 1am,
whispering "At least they will celebrate at ITP tomorrow. All attendants of
the Quantum Hall program will celebrate the Physics Prize". I responded
"you are early" and he admitted that he could not force himself to stay up
the additional 3 hours and wait for the announcement of the Chemistry Prize,
as he had done the previous year. He was just too tired. We are Swedes and
the Nobel Prize events have been an integral part of our lives since we were
small children. But what we in Sweden could follow by TV and radio during
day time had to be checked out on the internet during the night, now that we
lived in Santa Barbara, 9 time-zones away. So it came to be, that the most
important Chemistry Prize in our lives was unnoticed by us until 8am when
my mother, Marianne, forced by my father Ingemar (who was tired of all
speculations), finally phoned us and asked if it was really my Professor that
had been awarded the Chemistry Prize.

And yes it was! Walter Kohn was indeed awarded the Nobel Prize in
Chemistry. Since July 1997 I had been a postdoc with Walter. It is a story in
itself how I switched from quantum field theory to DFT, but I will save it for
another time. Here, I'll share anecdotes that came to my mind when I leafed
through the folder with newspaper clips and memorabilia I have saved from
this time.

First in the folder is a front-page of the Santa Barbara News-Press of October 14, 1998. The headline is a quote from Walter: "You could say everything
is rosy." Since nobody at UCSB could reach Walter by phone, Alex Kamanev,
who was Walters other post-doc at that time, and I were sent by the staff at
ITP (Institute of Theoretical Physics) to ask Walter how he wanted everything to be set up for the celebration of his Prize. As usual, I had roller-bladed
to campus but Alex had, wisely enough, brought his car, parked free at Goleta
beach of course. He also suggested buying flowers for Walter and Mara. We
stopped at Trader Joe's on Milpas and encountered the first problem. How
to choose a bouquet appropriate for this kind of event? As a Swede I was

supposed to know, but I can tell you that not even in Sweden do they teach us that in school. After some discussion we simply decided on the most beautiful ones, a large bouquet of red roses. At the Nobel Banquet, the Blue Hall in Stockholm is certainly not decorated with red roses. However, we were in California, where they serve Champagne in plastic glasses, which is about as bad as it gets with European eyes. Most importantly, Walter did approve of our choice and this made the headline.

The day after the BIG day, we as usual had one of these now famous lunch meetings; Walter would absolutely not cancel this tradition. Again, we had an etiquette problem. Thomas and I brought Champagne, and ensured that we could toast in real glasses. ITP actually had a set of real glasses, although not crystal, hidden in a cupboard and I cleaned them thoroughly before the meeting in the 6th floor seminar room in Broida Hall. But what is the proper choice, according to Nobel etiquette, serving the Champagne before of after a scientific talk? Walter finally solved the problem for us by asking if the Champagne was just meant to be looked at or if we could perhaps taste some of it ... . At that point I decided to not bother about manners any more, we where in California and I decided to do it the Californian way, that is, relaxed.

Monday October 19, I celebrated my birthday and the same day Walter's and my first paper together was published in Physical Review Letters. I had experienced a week more exiting and wonderful than I can ever convey.

Now, let us jump forward in time. Because of sickness in the family Walter could not attend the Nobel festivities in December of 1998. However, relieving words came and Walter prepared to go to Stockholm and deliver his Nobel Lecture in January 1999 (a requirement for receiving the Prize money, Walter told us). That month, a most hectic part of my life had just started. In the summer of 1998 I had interviewed for a position as assistant professor at the Royal Institute of Technology in Stockholm (KTH). Somewhat surprisingly, I was offered the position and they wanted me to begin immediately. My suggestion, on the other hand, was to finish my 2-year post-doc with Walter and not start until fall 1999. After some negotiations we agreed on January 1st, 1999, as a starting date, however, I was allowed to work for extended periods of time at UCSB. I would spend January, March, and May at KTH and February, April, and the summer in Santa Barbara working with Walter. When I learned that Walter would give his Nobel lecture in January I could not believe in my good luck.

January 1999 was one of the coldest months on record in Sweden, newspapers wrote about power outages, freezing pipes, and broken cars. The day Walter would arrive to Stockholm was cold, very cold, colder than any of the previous days and I suddenly remembered one of our last conversations, when we were both still in sunny Southern California. Walter had asked what to wear in January in Sweden, he only had a raincoat with an inset, did I think that would be enough? I remembered answering "It depends" but for some reason we dropped the subject and I did not get to finish my advice. When remembering this I got concerned and phoned Professor Börje Johansson who was to pick Walter up at the airport, urging him to bring an extra coat in case Walter showed up in the raincoat. Of course, my worries were unfounded. Walter arrived from Vienna, were they also had cold weather and, of course, he knows how to dress. In addition, Walter charmed the Swedes by insisting on enjoying the cold weather. A headline in the Uppsala newspaper "Upsala Nya Tidning" reads "Professorn från Kalifornien njuter i kylan" ("The Professor from California enjoys the cold weather"). My lasting memories from the Nobel Lecture at the Royal Swedish Academy of Sciences itself are that Walter did not need more than the 45 minutes he was allotted and that I seldom have seen a greater number of elegantly dressed ladies at a ~~Physics~~ Chemistry seminar.

The next morning, I met with Walter at his hotel, the distinguished Grand Hôtel, for breakfast and handed over printouts of 200+ e-mails forwarded from Walters account in Vienna to me. After breakfast, a taxicab drove us to Uppsala and during the one-hour drive we actually managed to discuss some Physics. Walter gave a rerun of his Nobel Lecture at Uppsala University, the oldest university in Scandinavia, founded in 1477, this time in a more relaxed setting. The day was filled with meetings and we had a wonderful lunch at an old restaurant. Walter stayed in Uppsala while I took a train back to Stockholm. The next day I went back to Santa Barbara and my family.

We were still trying to work closely together during the spring of 1999, although we were seldom at the same place on earth at the same time. Accordingly, we took every opportunity to discuss our project. We had breakfast and lunch meetings and I took walks with Walter; at the APS March meeting in Atlanta, at a conference in Hindås in Sweden and sometimes in Santa Barbara.

On April 11, 1999, UCSB gave a dinner in honor of Walter and Mara. Friends and colleges of Walter got to give 1-3 minutes talks and I presented the above mentioned coat anecdote. However, much more importantly, Walter got a present, a helmet to wear when rollerblading. This reminds me about the delicate topic of our ongoing debate regarding rollerblading safety. I use all available safety gears, helmet is a must but I also use knee and elbow pads as well as wristguards. Walter used to laugh when he saw me leaving work looking like a football player. I have never seen Walter rollerblading but I knew he did, even before all newspapers featured pictures of him skating. To my horror, he was not even wearing a helmet! I tried to convince him that even if he thought that he did not need his full brain capacity any longer, many people appreciated his thinking immensely and did not want him to impair it by hitting his head. At this dinner, I even tried to convince Mara to put pressure on Walter to really USE the helmet we now knew he had. I'm not sure he ever did, but I know that he did not use any other safety gears. A few years later Walter broke his right wrist rollerblading at East Beach. I visited him at UCSB at that time and while learning to write with his left hand he assured me that his doctor agreed with him that wrist guards had not made a difference in this case. Anyway, he promised me to wear safety gadgets while rollerblading, at least helmet and wrist guards. History is not without irony, however, and in order to give a point back to Walter on that you cannot always prepare for the worst, I need to mention that in the summer of 2002 my daughter Carolina broke her wrist on East Beach. It happened while jumping on and off a surrey (4-wheeled bike with seats that you steer as a car). Who thought about using safety gears while pedaling along a California Beach in a surrey? It is not even mentioned on the waiver forms.

When Walter returned to Stockholm in December 1999 for his postponed Award Ceremony in Stockholm Music Hall and Nobel dinner at Stockholms stadshus, he continued to charm all Swedes. Among other things newspapers featured pictures of Walter ice skating in Kungsträdgården, the Central Park of Stockholm. Thomas and I had, to our great joy, gotten last minute invitations to the Nobel events. I bought a beautiful dress and Thomas rented tails. (I used the dress for the second time when visiting the Santa Fe Opera with Walter in connection to a workshop I organized in Albuquerque in August 2002.) At the Nobel dinner, with the King and Queen of Sweden, it was time for me to worry about Nobel etiquette again. Fortunately, I was seated together with experienced Nobel dinner participants, Professors of the Royal Swedish Academy of Sciences and their spouses (married couples are not seated together, though, that is against the etiquette). They instructed me what is OK to bring home as souvenirs (chocolate coins, match boxes, your own seating card) and what is not (glasses, cutlery, the Kings seating card). They taught me how to get the napkin to stay in the lap on a dress without friction (the solution is clips, which they had brought a few extra of for inexperienced people like me). Since I had mentioned my children they arranged for me to bring home one of the candy "N"s from the top of the dessert ice cream. I could write several more pages about this event, for example, how do you seat 1270 persons at 65 tables? You give each guest a 72-page booklet (nice souvenir) with a foldout map over the table setting and an alphabetic directory with a code indicating the table and seat assigned. I had code 21:14, Thomas 61:21, and the US ambassador to Sweden A:32. The guests of honor at the A table came marching in together with the King and the Queen. Walter was seated with Lena Hjelm-Wallén, the vice Prime Minister of Sweden (at that time), only two seats away from the King himself. I could tell about all famous people, the dancing, the menu, and the artists that entertained us. But even though all this was spectacular my lasting memory is the pride Walter showed when he received the Nobel medallion and diploma from the hands of Carl XVI Gustaf, the King of Sweden, earlier that day. At the web page of the Nobel Foundation, I later found a photo of Walter, taken at that moment. I downloaded it and use it as a background picture on my job computer. It serves two purposes: it gives me inspiration to do long-term research and it reminds me that just being a researcher is not enough, I also need to be a human being with compassion and ethics. In both respects Walter is a role model for others and me. "Walter Kohn: A Nobelist with a Heart", "Of all the Nobel Prize winners I've known he's by far the nicest" are only two quotes in my memory folder illustrating this. With the hope that everyone that reads this will feel the same gratitude and inspiration as I do, I conclude with the photo.

Proud Walter Kohn receiving the Nobel Prize, Stockholm, December 1999 (© 1999 The Nobel Foundation, photo: Hans Mehlin)

**About the Author**: Ann Mattsson is a physicist in the area of condensed matter at Sandia National Laboratories in Albuquerque, NM. Her field of research is since 1997, when she was a post-doc with Walter Kohn at UCSB, Density Functional Theory. She develops new exchange correlation functionals. Contact: Surface and Interface Science Dept., MS 1415, Sandia National Laboratories, Albuquerque, NM 87185-1415, U.S.A.; aematts@sandia.gov, www.cs.sandia.gov/dft/index.htm

# Walter's Contagious Conscience

Eric McFarland

University of California, Santa Barbara, U.S.A.

Walter and Mara participating in the Santa Barbara Ad Hoc Committee Opposing Unilateral Preventive War in Iraq. January 2003

For many of us, Walter continues to exemplify the notion of immortality through mentorship. Often these are lessons on life. At the many Kohn family barbecues, Walter and Mara bring together friends, family, students, and visiting scientists for lively discussions ranging from density functional theory, religion, and the nuclear test ban to what possessed the giant swarm of bees to appear so suddenly and drive us into the house away from their spectacular patio overlooking the Santa Barbara harbor. Here Walter and Mara showed how one's students, colleagues, friends, family, professional work, and societal responsibility were each independently important yet inseparable, like breathing and eating; neglecting one brings down the entire system. It was here that I came to know clearly that the question of sacrificing one's family, friends, or societal responsibilities for our scientific work would never need to be addressed. Walter showed by example that it was not the right question to ask.

Eric McFarland

As a recently hired junior faculty member, I was drawn to his theoretical work on electronic phenomena associated with gas-surface interactions; however, it was during our work together attempting to end the inappropriate management by the University of California of the U.S. Nuclear Weapon Laboratories that our friendship developed around common beliefs. I'm not so sure how effective or helpful I was in the process, however, Walter's deeply held belief that every scientist, every citizen, has an obligation and responsibility to act upon their social and political conscience impacted me greatly. The example he sets in this regard is contagious and has widely infected those of us who have had the privilege to work with him. If Walter Kohn could take the time to get involved, we certainly could as well. It was Walter's example, advice, and encouragement (mentorship) that helped motivate future work for me in Kosovo and Mexico, and that I have tried to pass on to my own students and colleagues who have subsequently taken time out of their own busy lives to get involved in socially important causes. Most recently, as our nation began what appeared to be an aggressive unilateral military campaign in the Middle East, Walter's call to my office was a gentle reminder to raise my head above the distractions of an overflowing desk to issues of far greater importance. At Walter's urging, we called to action a local Ad Hoc Committee and created together a statement of opposition to unilateral preventive war against Iraq. Leading by example, Walter's hard work and dedication was remarkable and in short order he had rallied groups of Nobel Laureates, physicians, religious leaders, and business executives to support a responsible call for our government to seek broad international support before committing the world to further violence.

With profound respect and admiration, and in friendship,
Happy Birthday, Walter

**About the Author**: Eric McFarland is a professor at the University of California, Santa Barbara. His research concerns the engineering and chemical physics of catalysis. Current Interests include chemically induced electronic excitations in metal surfaces, nanocluster catalysis and photocatalysis. Contact: Department of Chemical Engineering, Engineering II, University of California, Santa Barbara, CA 93106, U.S.A.; mcfar@engineering.ucsb.edu

# Memorable Moments with Walter Kohn

N. David Mermin

Cornell University, Ithaca, U.S.A.

Just 40 years ago I was applying for academic positions from Birmingham England where I was finishing two postdoctoral years in Rudi Peierls' wonderful department. In those early post-sputnik halcyon days you then applied for assistant professorships and collected the offers that came in by return mail. (I exaggerate only slightly.) Peierls had suggested that I send such a letter to Walter Kohn, who disappointingly offered only a two-year postdoc. But I had never been to California and Peierls clearly had a high regard for the man, so this otherwise noncompetitive proposition was still very tempting. I asked the people at Cornell, who did come through with a faculty position, whether I could defer it for two years to do a postdoc in La Jolla. Two, they said, was too many but one would be OK. Walter said one was fine with him, so in August 1963 I showed up in La Jolla.

Walter was not there. He was finishing a sabbatical in Paris. But there were several other terrific postdocs (Vittorio Celli, Bob Griffiths, Lu Sham), the Physics Department was still located right on the beach, Dorothy and I found a fantastic house in Del Mar on a cliff overlooking the Pacific for $128 a month, and life was good. (I had written a letter from Birmingham to an acquaintance from college who was then a postdoc in La Jolla, asking what it was like there. All I remember from his reply was "Volley ball is standard on the beach at noon.")

Eventually the moment of truth arrived. The boss returned. The holiday threatened to end. Walter invited me to his office to say hello. It was immediately clear that this was a very kind, charming, witty man. After we had exchanged pleasantries, he told me about a little theorem he and Pierre Hohenberg had proved back in Paris. The proof was one of those clever three-line arguments that wouldn't have occurred to me if I had thought about it for a hundred years, but was utterly simple and transparent when Walter laid it out in front of me.

He asked me to think about how to generalize the theorem from the ground state to thermal equilibrium. I returned to my office to consider it and quickly realized that a strange variational principle for the free energy that I had formulated in Birmingham for an utterly unrelated purpose, seemed to be tailor made for generalizing the Hohenberg–Kohn theorem to non-zero temperature. It took me less than an hour to check that their proof did indeed go through in exactly the same way if the ground state variational principle they used was replaced with my thermal equilibrium variational principle.

So I went back to Walter's office and knocked on the door. Here's how you do it, I said. He seemed somewhat taken aback by this and before I got very far into my explanation he kindly offered to explain again what the problem actually was that he wanted me to work on. He treated me to his beautiful and transparent argument again. I said yes, that was what I had understood the argument to be (he really had explained it very well the first time) and my point was that it worked just as well when the temperature was not zero if you used this variational principle I was trying to tell him about. He was deeply skeptical. Slowly it began to dawn on me that this was the problem he had hired me to spend the year working on.

It took me a day to convince him that I had indeed answered his question. Then he was very pleased and I, needless to say, was ecstatic. Throughout childhood I was the last to be picked when baseball teams were being formed. I could never get my bat to make contact with the ball. But one day, by sheer chance, I got the bat in the right place at the right time and the ball went sailing over the heads of the outfielders. I must have been ten years old. It was a magical moment. Now I was 28, at the beginning of my career, and it had happened to me again. Never in the forty years of professional life that followed did I ever again have as glorious an experience.

Since I had finished the year's work, Walter encouraged me to think about whatever I felt like thinking about – pelicans, whales, body surfing, physics. We became good friends. As we said goodbye at the end of the year he said, "By the way, when you get to Cornell, write up that theorem. Some day it may be important."

Frankly, I wasn't so sure. But I did feel he ought to get something out of having maintained me in that semi-tropical paradise for a whole year, so I dutifully wrote a very short paper in Ithaca. On the occasion of writing this memoir I subjected my Hohenberg–Kohn Corollary to a citation search. The first decade after it appeared (1965–1974) bears out my doubts. The number of citations per year ranged from 0 to 2, most of them traceable back to Walter and his collaborators. Then, ever so slowly, Walter's "some day" started to dawn. During the next six years (1975–1980) annual citations varied between 3 and 7. Between 1981 and 1993 there were between 17 and 23 citations per year (except for a 12 in 1982). And from 1994 to 2002 (with the Nobel year 1998 smack in the middle) there were never less than 28 and as many as 40. It may now be my second most cited paper. Not bad for an hour's work! In 1998 I realized that it had also became my second publication in the field of chemistry.

Honesty compels me to acknowledge that Walter has a different view of this history. He maintains that it took me 24 hours, not just one, to do his postdoctoral project. Although his ability to recall ancient events is normally phenomenal, in this case I am rather sure that doing the job took an hour; it took me the other 23 to convince him that I really had done it.

Walter was not only directly responsible for my finest hour, but he was also present at my finest half-hour, a quarter of a century later. In 1989 I gave a lecture in St. Louis on quantum nonlocality at a March APS meeting with the nonstandard title "Can You Help the Mets by Watching on TV?" To my amazement and the astonishment of several of my friends, 2000 people showed up. (It was in a hotel ballroom but there was standing room only.) Among them, to my surprise and delight, was Walter, who had watched, bemused, as my interest in foundations of quantum mechanics slowly developed during the 1980's. Every time I met him during those years he'd want to know just what it was that bothered me about good old quantum mechanics. He was always very nice about it, saying he just didn't get it – what exactly was the problem? I knew that deep in his heart, though he was much too kind to come out with it directly, he didn't really think this was a fit preoccupation for one who had once been capable of doing a year's work in a day. (No, Walter, in an hour!)

After a wild half-hour talk and an even wilder question period, Walter made his way through clusters of fiercely arguing people up to the podium to say hello. He shook my hand warmly, beamed his wonderful smile at me, and said "I still don't get it."

I cannot claim to have been present at any of Walter's finest moments, but I was there for two quite fine ones. The first was his 60th birthday celebration in Santa Barbara, twenty years ago. Vinay Ambegaokar and I sang a long song, setting his CV to music in the form of a well-known Gilbert and Sullivan number. The banquet audience, provided with copies of the words, served as chorus. Well do I remember the voice of Pierre Hohenberg, who served as master of ceremonies, lustily belting it out.

Unfortunately Vinay and I have both lost the text, and when I wrote Walter a few years ago asking for a copy of the copy I gave him, he had to confess that he couldn't find it either. [If any readers of this memoir have kept their libretti, please do send me a copy!] I have wracked my brain to recreate the words for this occasion, but my brain has fought back and all I can offer for this 80th birthday Festschrift are fragments. The recurrent refrain was "Now I am director of the ITP" so the whole business would not work, now that it has become the KITP.

I can recollect the opening verse, which celebrated Walter's very first publication:

> When I was a lad I thought a lot
> About the heavy and symmetric top.
> I thought so much they published me
> In the 'Merican Mathematical Society. . . .

N. David Mermin

But aside from that I remember only the verse on Kohn anomalies, the second
line of which was inspired by a wonderful phrase from Walter's paper:

> Amongst the phonons I could see
> The image of the surface of the Fermi sea!
> [*Chorus:*] Amongst the phonons who but he
> Could dare to see the image of the Fermi sea?

After the performance Walter remarked that he had always been quite proud
of that line, as indeed he should be. I have always admired him as a prose
stylist as much as I admire him as a physicist – whoops, chemist!

I was also present, along with about five thousand others, at Walter's meeting with Pope John Paul that he spoke about so movingly in a subsequent
talk at Santa Barbara[1]. To appreciate the moment from my perspective you
have to know the old joke about Louie:

> Louie knew everybody. "There ain't nobody I don't know," he
> boasted. "C'mon," said Al, "you don't know the Pope." "Want to
> bet?" said Louie. So off they flew to Rome, where we find them
> standing amidst a huge throng in Piazza San Pietro, waiting for the
> papal blessing. People in the crowd who walk past them are saying
> "Hi Louie," "Wie geht's? Louie," "Ciao, Louie," etc. "Stay here,"
> says Louie to Al, and disappears into the crowd. Some time later
> the Swiss Guard appear on the steps, there's a fanfare of trumpets,
> and out comes the Pope onto the balcony of St. Peters, arm in arm
> with Louie! An enormous cheer goes up from the crowd and a monk,
> standing near Al, turns to him and shouts over the roar, "Who's that
> guy up there with Louie?"

At the end of a week of sixty simultaneous conferences in all areas of human
knowledge, in celebration of the Jubilee Year 2000, there was a final superplenary session at the Vatican. A group of us walked there from our hotel
and just as we came upon an enormous array of outdoor tables alongside
St. Peter's, loaded with five thousand little plastic cups of espresso ("With
the compliments of His Holiness") we noticed that Walter had disappeared.
There were rumors that he had been siphoned off for something special, but
nobody knew for sure. The rest of us were conducted to seats fifteen or twenty
rows back from the stage in an auditorium that made my ballroom in St. Louis
look like an intimate seminar room. After a couple of hours of inspirational
warm-ups, His Holiness appeared on the stage, listened to four people each
give a one minute summary of 15 conferences, and then gave some remarks of
his own to put it all into perspective.

---

[1] (see http://www.srhe.ucsb.edu/lectures/text/kohnText.html)

That being done, a couple of dozen cardinals formed a line in the central aisle and one by one walked up onto the stage and, kneeling before the Pope, kissed the ring and were blessed. They were followed by an even larger collection of Archbishops, who were followed by an endless string of Bishops. Then the laity joined the procession and suddenly we noticed the back of a familiar head slowly approaching the stage. Could it be our missing Walter? What would he do? Nervously we awaited the meeting. Yes, it was Walter! When the time arrived, standing before the Pope, he shook the hand and launched into what, in comparison to the preceding brief encounters, could only be called a little chat. At this precise moment Giovanni Bachelet, a native Roman, who had encouraged several of us, including Walter, to participate in this extraordinary week and who had made it clear to us that he was a devout member of the Church, called out to me across several aisles: "David, who's that guy up there with Walter?"

My mind has just favored me with the final set of verses from our 60th birthday song. They were delivered in the form of an encore after everyone thought it was all over, and they make an appropriately festive note on which to end this 80th birthday offering:

> Happy Birthday, Walter Kohn;
> I bet you thought our little song was done,
> But like your own career, you see,
> It turns about most unexpectedly:
>
> From a lad with a top you have grown to be
> The Founding Director of the ITP.
> *[Chorus:]* That boy with a top has gone on to be
> The one and only leader of the ITP.

Were the party today, I would change the final line to

> A certified Practitioner of Chemistry.

**About the Author**: N. David Mermin is a professor of physics at Cornell University, a member and former director of Cornell's Laboratory of Atomic and Solid State Physics, and a founding member of Cornell's Faculty of Computation and Information. He has done research in statistical physics, low temperature physics, mathematical crystallography, and foundations of quantum mechanics. He is currently interested in the theory of quantum computation and its implications for quantum foundational issues. Contact: Physics Department, Clark Hall, Cornell University, Ithaca, NY 14853-2501, U.S.A.; ndm4@cornell.edu; www.lassp.cornell.edu/lassp_data/NMermin.html

# Impressions of Walter Kohn

Horia Metiu

University of California, Santa Barbara, U.S.A.

One month after I arrived in the US, I was sure that I could write a book about America and the Americans. I thought that I understood clearly and vividly their national character and what makes them different from Romanians. After living in the country for thirty years, too much exposure has dissolved the Americans into a mass of thousands of shapeless details. The same thing happened with my knowledge of Walter Kohn. Having seen so much (but not enough) of him, my impressions lack relief and are softened by too many nuances. As I try to write about him, my prose resembles that of Niels Bohr: one statement and ten secondary, qualifying sentences. To lighten the burden I decided to abandon any pretense of describing or explaining Walter Kohn. Here are a few stories and thoughts stimulated by Matthias's request.

There is a story that Walter likes to tell, which illustrates his way of doing Physics. It involves the motion of a pendulum and I am telling you what I remember that Walter remembers. This is a dubious transmission line and the information may have been distorted. Please do not use this material to learn about the pendulum; it is a lesson about Walter. While an undergraduate at the University of Toronto he took Mechanics from a Russian professor, who devoted some time to the pendulum. This is not the pendulum in your grandfather's clock, but a weight suspended on a string. If you give the weight a kick, it will rotate and move up and down in a fairly complicated motion. The Professor posed the following problem: if you follow the height of the weight, you find that from time to time it reaches minima at certain angles; calculate these angles and find out what controls their magnitude.

This is one of those questions in "fundamental" Physics whose importance is hard to determine. It is obviously of interest to those who have to stand under a moving pendulum and want to avoid being hit in the head. This is also the kind of qualitative question that gets Walter's heart racing.

When he gets to this point of the story, Walter's eyes begin to sparkle and the wrinkles around his eyes are activated. You can tell that he anticipates the pendulum to reveal miracles. The Professor had some trouble with this problem and the class was in suspense: Is the pendulum winning, or will rationality prevail? Following an appropriate pause, to let the gravity of the moment sink in, the Professor announced that he is going into the complex plane. After over sixty years, Walter can still relive the moment and his amazement over this trip away from the safety of the real axis, the seat of all physical quantities. Then the Professor revealed, clearly and incontrovertibly, that the complex plane had a singularity that controls the behavior of the pendulum. Walter finds this fact extremely beautiful and significant; it gave him enormous pleasure to hear about it and it is a memory that still delights him. He claims that this was one of the revelations that turned him towards theoretical physics.

This story is significant. For most physicists a pendulum is a dumb and trivial thing. If they were forced to say something nice about it, they would probably mention how amazing it is that such a complicated motion is completely predicted by Newton's equation. While important, the agreement with experiment is not pregnant with consequences. The role of the pole in the complex plane and its power to influence the motion on the real axis is deeper and more fruitful. Resonances are poles of the $S$-matrix, the decay length of a Wannier function (a topic dear to Walter) is controlled by a pole, poles in the Green's function give the energies and the lifetimes of quasi-particles, semi-classical mechanics lets trajectories go in the complex plane, a branch cut in the Green's function introduces true irreversibility. Like politicians in "smoked-filled rooms", the singularities in the complex plane manipulate reality. In focusing on the pole, Walter's taste led him directly to the true essence of the phenomenon. He does this often and he does it well.

Gibbs once said that he had done little in Thermodynamics except find the point of view from which the physics appears in the simplest way. Walter loves simplicity and clarity as much as Gibbs, but his fascination with the pole tells a different story: he also loves the point of view from which a phenomenon is most mysterious and beautiful. He is not a mere equation cruncher, and intellectual elegance matters to him. He measures theoretical results by the emotion they produce as well as their utility.

I am intrigued by the fact that Walter never seems tired or exhausted. Most of my colleagues look, most of the time, on the verge of exhaustion. I am not an exception, and everything seems boring, lifeless, and extremely difficult. I have never seen Walter in this state. He is always lively and animated. It is too bad that the word gay, as in "gay Paris", has been so thoroughly corrupted, because it would describe well Walter's spirited and light-hearted attitude. I asked once what his secret was and why he is never tired like everybody else I know. He smiled very mysteriously and promised that he will tell me sometime. He never did, and left me to guess for myself. We are all busy trying to do good science and most people I know attack the problem they are working on. The term is not accidental: they do go to war. Their mood evokes trenches, mud, blood, tanks, and artillery. Life is suspended until the battle is won or lost. This is not Walter's approach. He seduces a problem. He questions Nature but does not torture it. Seduction is a merry and subtle business. Nobody has seen an exhausted seducer, because working ten times as hard at seduction does not increase the rate of success tenfold. On the contrary, it reduces it. Finesse and imagination are more important than persistence and sweat.

There is a wonderful sentence by a Japanese artist who followed Einstein during his trip to Japan, in an attempt to capture his spirit in his drawings. He said: "Professor Einstein speaks softly as if he is afraid that he might scare the truth". Professor Kohn does research in that way.

While he is extremely curious about Physics, and it is clear that he derives great pleasure from learning about an interesting problem, Walter chooses very carefully what he works on. There is a filter that will not let him get engaged, unless the problem deserves his attention. He is not besieged by an urge to work, to keep busy. He does not feel that he must have a large group, get a lot of funding, or be constantly in motion. He is not like a shark, which cannot breathe unless it is in motion. He has not written many papers (by the standards of most theoretical chemists) but they are all clever, elegant, useful, and often important. Walter is proof that quality may be inversely proportional to volume. I am not saying that we would all be Walter Kohn, had we written fewer papers. But, maybe if we did that, we might have approached the best quality we are capable of.

Walter has not had an easy life. He grew up when Austria was seized by a rabid and disturbing anti-Semitism. He had to go to a school for Jews, created to ensure that Austrians were not contaminated by contact with the likes of Walter Kohn. He escaped just in time to England, where Scotland Yard established that the young boy was not a threat to the stability of the British Empire. They shipped him to Canada where he became an enemy alien. He was a Jew in Austria, but an enemy German in Canada, since some government body decided that there is a high probability that fugitive Jews will spy for Nazi Germany. His battle for entrance at the University of Toronto deserves mention. He and a few other Jewish refugees lacked papers and proper credentials and this is a severe handicap in an encounter with any bureaucracy. In addition, the Chairman of the Chemistry Department was not going to allow enemy aliens in his building. Without chemistry courses and laboratory, a degree in physics or any other science was not possible. Walter may have become a distinguished story teller.

Providence interfered through a Dean who decided that no rules and opposition should stop these boys from going to school. He arranged meetings for Walter with the Chairman and with the President of the university, perhaps hoping that personal contact would make the aliens seems less dangerous. These meetings are among the very few occasions when Walter's charm did not give the desired results. Neither the Chairman nor the President was moved enough to make a favorable decision. When these visits failed, the Dean declared the whole bunch "special students". This placed them at an advantage since such a designation was new and there was no established policy for dealing with them. The Dean was now free to set the rules for "special students" and he declared them admitted. That was acceptable to all parties and pretty soon Walter and the others were allowed to tackle mathematics and even chemistry. Those of you who follow politics will recognize the method: When an idea encounters resistance, change its name.

The same Dean, who was teaching calculus, noticed that the special students were a bit rusty, so he tutored them on Saturday afternoon, until they caught up with the rest of the class.

It is typical of Walter that he remembers with reverence and admiration the Dean and has no rancor when he mentions the Chairman. The other day he even praised that man's good work on nucleation. Walter knows that bitterness harms more the one who bears it, than the one who deserves it. It is remarkable how well he follows his own counsel, something that most of us have difficulty doing.

There is a famous story about a Pauli letter of recommendation for Löwdin. It said: "I have known Löwdin for two years and I have nothing bad to say about him". The letter puzzled the recruitment committee, until Pauli insiders explained that he had something bad to say about everybody and that such a letter was a ringing endorsement. Löwdin was hired, or so the story says. A letter like that from Walter Kohn would be worthless. In twenty years I have never heard him say anything bad about anyone. In fact he dislikes hearing slander, even when it is deserved and it is funny. I don't know if this inclination to kindness is innate or is cultivated, but it is striking. I noticed that he was reading recently a book on Buddhism. He will have little difficulty in becoming an exemplary Buddhist monk.

Having lived for thirty years under and among communists I developed a sturdy suspicion of any change that is mostly based on good intentions. This does not make me a good liberal. Walter, on the other hand, is, in words and deeds, one of the most liberal men I know. In academia, opposing liberalism is not worse than a crime, but it is harder to forgive. A liberal in academia is like an infidel in Mecca: You need to watch what you say, if you value or need a certain degree of tranquility. One would think that after living for thirty years under communism, this should not be too difficult for me. But, occasionally I forget myself and I defend a corporation, or the drug industry, or a budget cut, or a tax cut, or I decry a tax increase. My statements cause amazement and horror: the caveman in the Faculty Club. Let's put him on the menu: the Conservative Faculty Burger. I have never received such treatment from Walter, who is a committed liberal who practices his creed. If I say something controversial he wants to know my reasons. He will discuss them without questioning my morality, or exhibiting his. He will accept the fact that political life is complicated because we are all likely to grasp at half truths. He knows that the role of political discussion is to establish the range of acceptable opinion, rather than exhibit moral superiority. We disagree often but he is satisfied if he thinks that my position is defensible, even though he opposes it.

Walter is one of the most tolerant persons I have met. And strangely enough, this tolerance gives his opinions more weight. I talk to him, I think that I made my case rather well, but a few days later I find that I have drifted closer to his position.

Finally, there is the Nobel prize. A halo is placed around the laureate's head and life is no longer the same. My secretary used to tell me that Walter called, by stacking his message along with all the others. After the prize, the largest yellow stick-it made by 3M will be glued in the middle of my computer screen, with the largest and most emphatic letters: Call Walter Kohn. The post-Nobel calls were mentioned with reverence. There was a touch of pride also: not all our callers are mere mortals. Even the University noticed Walter. He got a full-time secretary and an additional, small office for visitors. He was no longer in danger that one of the stacks of books piling to the ceiling of his tiny office would fall on his head. Now that head had become an asset, and had to be protected.

It is amazing to see the extraordinary discontinuity that the prize causes, even though nothing really changed. Instead of the French "the more it changes, the more it stays the same", we have here "nothing changed and nothing is ever the same". The man awakened by the call from Stockholm, is the same man that went to bed the evening before. I am sure that he was a bit happier, but the call did not change his knowledge of physics or his ability to teach it.

Everything changed, except Walter. I watched him carefully since I thought that some aspect of human comedy will be revealed to me. That I would witness, with anticipated sadness I must say, Walter's transformation into a Nobelist. There was absolutely no visible change! Walter knew who he was, he liked who he was, and that would not be changed by a phone call. I am very glad, because I liked very much what Walter was and still is. I am glad that the University changed and I am glad for his office and secretary.

I warned you that I will just tell you a few stories about Walter, and that they might appear aimless. The point I was trying to make with them is that Walter belongs to a very rare species: a fully civilized man. If most people would be like him there would be no war, no poverty, and no cruelty and rationality and kindness would rule the world.

**About the Author**: Horia Metiu is professor of chemistry and physics at the University of California, Santa Barbara. In his research he is trying to find interesting problems that are amenable to a theoretical solution. He is currently working on the mechanism of catalytic reactions and on using theory to find better thermoelectric materials. Both activities rely on density functional theory. In the past he has worked on surface enhanced spectroscopy, Penning spectroscopy, rate theory with applications to surface science, femto-chemistry, theory of spectroscopy in time domain, dynamics of electrons in quantum wells, protein folding, single-molecule quantum dynamics and crystal growth. He is finishing a textbook on physical chemistry that should be published next year. Contact: Department of Chemistry & Biochemistry, University of California, Santa Barbara, CA 93106-9510, U.S.A.; metiu@chem.ucsb.edu

# ... Small Matters

Douglas L. Mills

University of California, Irvine, U.S.A.

I must begin these remarks with a tribute, though this was not in the request I received. Like many others, I have been reading Walter Kohn's seminal papers on fundamental aspects of condensed matter physics since my graduate student days. The papers are all most notable for their clarity and the simplicity of the mathematics one encounters. On many occasions, after reading through the material, I found myself saying something like "of course things go that way, I could have written this myself". However, I didn't write the paper or think of the core idea. It is the case that the most important and fundamental new ideas and concepts in our field are very simple and obvious, once they have been set forth for the first time. I am reminded of remarks I have read recently in an essay by Steven Weinberg, who states that the very important and fundamental papers in physics are notable for their clarity. The new ideas are applied quickly because of this.

In the many lectures I have heard over the years, and in my personal interactions with Walter, what is most striking and unique is the careful, thoughtful and precise way he addresses almost any issue, from those in physics to those regarding other aspects of our life or our culture. One learns to listen carefully, and particular sentences remain fixed in one's mind. His comments are also interspersed with a delightful dry humor.

Various small incidents illustrate his approach, in my mind. For example, I attended a conference a few years ago on aspects of density functional theory, and Walter was in the audience. A very aggressive younger speaker was describing a very complex, technical calculation of the electronic structure of a complicated molecule adsorbed on a crystal surface. In the course of the analysis, the Lindhardt function entered the discussion. The speaker showed a plot of the Lindhardt function, which indeed looked quite smooth to the eye on the scale used. The speaker argued that this was a complicated but nonetheless quite boring mathematical function of no fundamental interest. As a consequence, no harm was done by approximating it through use of a simple polynomial form; to do so led to some simplifications in this particular analysis.

165

At this point, Walter raised his hand, and commented in his careful and deliberate manner that if one took the time to look at the function a bit more closely, in fact its structure was of some interest. The speaker put him down brusquely, and rolled on. So much for Kohn anomalies, charge density oscillations around impurities as discussed many years ago by Vosko and Kohn, and many other fascinating phenomena in metals.

In regard to his humor, quite a number of years ago, I was standing with him for a few minutes at a March APS meeting when someone I knew approached us and joined the conversation. I introduced him to Walter, and he remarked "Oh, I know who you are – I recognize you from your papers". Walter commented that this was very strange indeed, since he didn't think he looked like his papers. This fellow didn't quite know what to say next.

Finally, I had reason to telephone Walter about a matter a few days after he received the Nobel Prize in Chemistry. Perhaps like others, I was very struck not by the fact that he received the Nobel Prize, an award much overdue in my mind and that of many others I know, but by the fact that he received it in chemistry rather than physics. As I look through the list of eminent theorists in our field, Walter is perhaps the deepest intellect of the group, thinking always of fundamental principles, with specific applications a means of illustrating consequences of fundamental points. So I asked him his reaction to receiving the Noble Prize in chemistry rather than physics. His answer was that he deeply appreciated the welcome into this new community, and was eager for new interactions that this would produce. He then remarked that in his mind, the physics community was not sufficiently open minded and flexible enough to welcome the contributions of outsiders in this manner.

The matters I mentioned above all seem small, but as I remarked above, Walter has a remarkable talent for coming up with single, simple sentences, usually very pointed and sometimes humorous, and these are remembered for a very long time.

**About the Author**: D.L. Mills is Professor of Physics at the University of California, Irvine, and serves also as Director of the Institute for Surface and Interface Science. He is a condensed matter theorist whose interests have centered on the physics of surfaces, interfaces and thin film structures, and his research has examined the theory of electron probes of surfaces, linear and non-linear optical interactions in this environment, and most recently his research has been focused on theoretical issues in the magnetism of nanoscale structures. Contact: Department of Physics and Astronomy, University of California, Irvine, CA 92697, U.S.A.; dlmills@uci.edu

# Walter Kohn in Cracow

Roman F. Nalewajski

Jagiellonian University, Cracow, Poland

Although my interest in DFT had already started in late seventies, when I was exposed to it during an extended stay in Bob Parr's Group at the University of North Carolina at Chapel Hill, my first personal contact with Walter Kohn came about fifteen years later, at the NATO ASI held in Italy (Il Ciocco) in 1993. I was then at the final stage of shaping the program of the DFT Symposium to be held in Poland's historic capital city Cracow next year, to commemorate the 30th anniversary of the modern DFT. The Il Ciocco conference was for me a great opportunity to both meet the theory inventor and to publicize the coming Cracow event among theoretical physicists and chemists.

The opening lecture by the "Father" of DFT, a good-natured, illuminating, "common-sense" style of his discussing sometimes difficult and controversial elements of presentations by other participants, and my personal conversations with him, limited not only to science but including also some personal recollections and reflections on science, art and history, presented in a witty and vivid way, had then left me under a strong impression of his wonderful, congenial personality, which was only strengthened by many future encounters. Personally, I always appreciate that in his talks Walter Kohn clearly identifies not only the specific successes of the theory and the promising directions for future work, but also the problem areas, which still await a more satisfactory solution, and hints upon possible ways of dealing with them. His inquisitive and open, sympathetic attitude towards other fields, following Eugene Wigner's unifying statement that the understanding in science requires understanding from *several different points of view*, is manifested in Walter's genuine interest in many numerical and conceptual applications of DFT in chemistry, which added decisively to the enormous impact that DFT has reached in the theoretical community, as indeed recognized in 1998 when Walter Kohn shared the Nobel Prize in Chemistry with John Pople. I recall, that a returning motive of our discussions was, and still is, why do chemists care so much for such unobservable concepts of traditional chemistry as molecular subsystems, e.g.,

At the Collegium Maius, the oldest part of the Jagiellonian University, June 1994

At the symposium venue in the Aula of the Chemistry Department, Cracow, June 1994

bonded atoms interacting substrates, localized chemical bonds, etc., and how do they approach the problem of defining them objectively.

Before the 1994 Cracow DFT conference, Walter Kohn had spent a day of reflection in the Auschwitz extermination camp, besides some sightseeing in the company of Ewa and Krzysztof Brocławiks. In his opening lecture He kindly took time to acknowledge some of his personal "links" to the Jagiellonian University through his Professors, Leopold Infeld and Fritz Rothberger, and colleagues, Marc Kac and Roman Smoluchowski, the son of Marian Smolu-chowski, who all had their scientific roots in the Jagiellonian University. I've heard more of Walter Kohn personal recollections of Leopold Infeld and his impressions of Infeld's memoirs during the DFT conference in Rome. Through-out the Cracow conference his active participation in discussions and informal exchanges has contributed decisively to the success of the symposium, alle-viating some of the apprehensions the quantum chemical community, firmly rooted in a tradition of the Hartree–Fock model and its refinements, originally had about the basic proposition of DFT. Here are three photographs of Walter Kohn from that meeting, including the one with his inseparable French beret, a habit – if I remember correctly – going back to the days of his collaboration with Pierre Hohenberg in Paris. My next brief encounter with Walter Kohn and his wife in Cracow was a few years later, when they visited the city during the annual Festival of Jewish Culture, on the occasion of the Documentary Photographic Exhibition of Mrs. Kohn's father.

At the university country estate at Modlnica, June 1994

**About the Author**: Roman F. Nalewajski is a professor at the Jagiellonian Univer-sity, Cracow, heading the Quantum Chemistry Group of the K. Guminski Depart-ment of Theoretical Chemistry. His research concerns conceptual and methodological aspects of quantum chemistry, and particularly the density functional theory, theory of electronic structure and chemical reactivity and catalysis. His current scientific interests include the use of the information theory in extracting chemical and "ther-modynamic" interpretations of the computed electron distributions in molecules. He is the author of the monograph on the "Charge Sensitivity Analysis" of molecular and reactive systems and a textbook on quantum chemistry (in Polish). Contact:

Roman F. Nalewajski

Faculty of Chemistry, Jagiellonian University, R. Ingardena 3, PL-30-060 Cracow, Poland; nalewajs@chemia.uj.edu.pl

# Walter Kohn

Venkatesh Narayanamurti

Harvard University, Cambridge, U.S.A.

I am honored to be considered a "close, personal" friend of Walter Kohn and to write about my recollections of him in honor of his 80th birthday.

I have been an admirer of Walter Kohn ever since I came to the United States as a graduate student in physics at Cornell. My first term paper assignment in a Solid State Physics course was on neutron scattering in lead and the structure in the spectrum due to "Kohn anomalies." When I went to Bell-Labs and subsequently became Director of Solid State Electronics Research, I began to learn from many of the great theorists who were there the enormous foundational impact Walter's work in "density functional" theory had on diverse problems in semiconductor physics – from surface reconstruction in silicon to the prediction of energy states of deep defect states in semiconductors.

So when I went to Santa Barbara as Dean of Engineering in 1992, and began to interact more closely with Walter, I discovered not only the physicist but someone who cared deeply about the broader implications of science on the human race. We even engaged in a "debate" at the local student-run radio station on the question of whether the University of California should continue to manage national laboratories such as Los Alamos because of their nuclear weapons mission.

After I moved to Harvard in 1998 and got involved in his nomination for the "Centennial" Medal of the Graduate School of Arts and Sciences, I learned about his joining the Physics Department as a Lehman Fellow soon after World War II. It was an especially exciting time and both faculty members and graduate students believed they were working close to scientific discoveries. Walter Kohn, I discovered, completed his graduate work in three years with a dissertation titled, "Collisions of light nuclei." It was certified by Julian Schwinger, Wendell H. Fury and Herbert Goldstein. The department chair, J.H. van Vleck and Professor Edward Purcell, besides Professor Schwinger became mentors and friends. Graduate student colleagues were Nicholaas Bloembergen, Robert Pound, Roy Glauber, Charles Slichter and at MIT, J.M. "Quin" Luttinger.

Dr. Kohn on the occasion of receiving the Centennial Medal, June 2001 (photo: M. Stewart; courtesy of the Graduate School of Arts and Sciences, Harvard University)

In June 2001 Walter was awarded the graduate school's Centennial Medal. The Dean's citation read:

"Bridging the gulf between domains of science, you have made seminal contributions to the fields of Physics and Chemistry and have provided in your own career a testament to the fundamental unity of human knowledge."

Walter, it is a great honor to know you and to be considered a personal friend.

**About the Author**: Venkatesh ("Venky") Narayanamurti is Dean of Engineering and Applied Sciences and the John A. and Elizabeth S. Armstrong Professor of Engineering and Applied Sciences at Harvard University. He is also a Professor in the Harvard Physics Department. He has published widely in the areas of low temperature physics, superconductivity, semiconductor electronics and photonics. He is currently active in the field of semiconductor nanostructures. Contact: Division of Engineering and Applied Sciences, Harvard University, Pierce Hall, 29 Oxford St., Cambridge, MA 02138, U.S.A.; venky@deas.harvard.edu

# A Class with Class

Herbert Neuhaus

Glenview, U.S.A.

It would be a pleasure to contribute a funny story concerning Walter and myself, and I am sure there would be one or two but since our association goes back 65 years and my memory unfortunately does not, I am at a loss at what to say. Of course, those were not funny or happy days after Hitler had occupied Austria in 1938 and we Jewish students were kicked out of school. I was allowed, however, as was Walter, to finish the fifth grade Realgymnasium (the ninth year in the public school system). We were informed that there might be a possibility to continue for another year in a Jewish school, though access was limited to Jewish honor students selected from all the schools in Vienna, especially those whose fathers had served in World War I. Since I qualified on both counts I was admitted to this school, the Chajes Gymnasium on Staudingergasse in the 20th district of Vienna. That is where I met Walter Kohn, who came from the Akademische Gymnasium. It was the most extraordinary school, the most extraordinary class and the most extraordinary atmosphere. Can you imagine, the very best 15 year old Jewish students of Vienna in one classroom! It was also the first time I experienced a coeducational setting (in Austria the sexes were, and probably still are, strictly separated below university level). There was a strong bond of comradeship, even friendship between students and professors and there was only one common goal: to learn as much and as fast as possible. The official curriculum was thrown out of the window and we pushed ahead up to and beyond university level. I remember that some of us went after school and after the school year had ended to the home of our mathematic teacher, Prof. Viktor Sabbath on Währingerstrasse to continue in higher mathematics – until he was deported to a death camp. Also, our physics teacher, Professor Emil Nohel, a former assistant to Einstein, made a great and lasting impression on us and I think, especially on Walter.

I do not know how many of our classmates survived the holocaust, but among the few who did, we can claim one Nobel laureate: Walter. Then there was my dear friend Rudi Ehrlich, who fled to Italy, accepted his Italian father's name and became Professor Rodolfo Permutti, taught Mathematics at the

Universities of Naples, Bari, Rome, then became chairman of the mathematics department at the University of Trieste. He is known for his "Permutti spaces" and other important mathematical discoveries (of course, I do not understand what I am talking about). Gertrude Ehrlich, no relation to Rudi, became Professor of Mathematics at the University of Maryland. Paul Sondhoff, an aeronautical engineer, worked for the United States Government, on projects so secret that his family would not permit him to associate with me after I had visited the Soviet Union in 1983. Karl Greger became Professor of Mathematics at the University of Göteborg in Sweden.

A (partial) class picture shows Walter in the third row with his arms around me (far right) and Paul Sondhoff (Bibi), Rudi Ehrlich-Permutti in front of me in the second row and Gertrude Ehrlich in front of Rudi, with Ilse Arnold-Levai next to her.

Has there ever been a class like this? I don't know how I fit into this illustrious group of scientists, I only made it to assistant professor of Medicine at the University of Illinois in Chicago. Unfortunately Rudi Permutti died last year, but Walter and I, independently, were able to visit with him in Trieste during the summer of 2001. Greger and Sondhoff died years ago.

Chajes Gymnasium in Vienna during the school year 1938/39, probably early 1939

Walter showed his love and respect for the memory of our teachers at the Chajes Gymnasium when he established and funded scholarships or prizes in honor of Professors Nohel and Sabbath in Human Rights and Natural Sciences and Mathematics for students at the Chajes and Akademisches Gymnasium. These prizes have been awarded now for a number of years.

When I heard of Walter's Nobel prize, I was so thrilled and proud as if I had won it myself! Also Austria, the country of his birth, the place that – after it became part of the German Reich – rejected him and drove him away and from where his parents were sent to their death – as well as the Austrian press, tried to claim him as their own ("at last a Nobel prize for Austria again"). After he received a letter and personal invitation from the Federal President of Austria, we had long and frequent telephone conversations which, I hope, were somewhat helpful in this moral and emotional dilemma and anguish.

Of course, it was not enough to congratulate him over the phone – I had the urge to shake his hand and embrace him. So, when he invited me to his home in Santa Barbara, I was more than happy to accept, and was fortunate to spend a wonderful and unforgettable weekend with him and Mara in their gorgeous home in the hills of Santa Barbara. After more than half a century he still remembered me playing Schubert for him on the piano! The picture shows the two of us in front of the University building, named in his honor. We were teenagers of 15 or 16 when we last saw each other – we were approaching 80 when we shook hands again!

Walter had the wonderful idea to arrange a reunion (the first and only) of this amazing 6th grade class (10th year of school education). We were able to locate five of our former classmates, of whom four attended the reunion

Santa Barbara in 1999

in Washington D.C. in May 2000. (Walter, Gertrude Ehrlich from Maryland, Ilse Arnold-Levai from Austria, and I from Chicago). Rodolfo Permutti could not attend because of poor health. I don't have to identify the individuals in the reunion photograph, having done so already for our Chajes picture and we did not age that much. Or did we?

Washington, D.C. in May 2000

Let me wish a very happy birthday and good health to our famous octogenarian and to all of us other octogenarians – we are all in the same boat, at least as far as age is concerned. May be there will be another reunion.

**About the Author**: Herbert Neuhaus was born in Vienna in 1923. After return from concentration camp he studied medicine in Vienna and graduated in 1950. Came to the US in 1951 and worked for the Department of Public Health and the University of Illinois since 1954, became director of the cardiopulmonary laboratory and then Medical Director and Superintendent of the Illinois Public Health Hospital in Chicago, also assistant professor of Medicine and assistant chief of staff at the University of Illinois Hospital. He is retired since May 1983. Contact: 521 Rio Vista, Glenview, IL 60025, U.S.A.; hb.neuhaus@juno.com

# Great Influences in a Small Country

Risto Nieminen

Helsinki University of Technology, Espoo, Finland

It is a great honor and pleasure for me to make a small contribution to Professor Walter Kohn's *Festschrift*. Unfortunately I have been able to meet him personally just a few times (probably he does not remember me). However, his work has been a guiding light for me as an aspiring scientist, and the few occasions when I have had the chance to meet him personally have left a lasting impression. His scientific work, which continues to flourish, has had a major influence not only on me but on the small but closeknit community of physicists and chemists in my country interested in electronic-structure calculations and related condensed-matter research. His personal warmth and utmost integrity as a citizen of the Earth are legendary. My younger colleagues know Professor Kohn only by name and by his papers, and cannot hide their envy when I can boast in the coffee table for having sometimes crossed paths with him.

I first came across Walter Kohn's name when I was a struggling undergraduate student at Helsinki University of Technology in the beginning of the 1970's. A group of experimentalists had hired me "to do theory" for them. They had made interesting discoveries of how plastic deformation influences the electronic properties of Al and Cu, having measured the positron lifetime spectra and electron momentum distribution (through angular correlation of annihilation radiation). They wanted me to calculate the expected changes in these due to dislocations and vacancies introduced in the deformation.

Initially, I tried to get started by consulting all sorts of books on the various many-body theories in fashion at the time. That did not lead to much. Eventually, I got help from two local many-body theorists, who suggested using Thomas-Fermi-type methods, which I did with limited success. Then I stumbled on a *Physical Review* paper by Lang and Kohn, dealing with the jellium model for metal surfaces. It was a major revelation. I remember having frenziedly gone through not only the Lang–Kohn papers but also their theoretical backbone, the Hohenberg–Kohn–Sham papers from a few years earlier. As far as I know, nobody in Finland had not paid any attention to these and electronic-structure calculations did not really exist. For me, this was the turning point.

A little later, as a graduate student in Cavendish Laboratory in Cambridge University, it was full gear ahead. Cavendish Laboratory in the mid-1970's was incredibly full of activity, and much of it was in electronic structure, thanks to Volker Heine and his colleagues. I was extremely fortunate for having been at the TCM group at that time, because a large fraction of leading electronic-structure theorists of my age group were there then as graduate students, post-docs or visitors. It was density-functional theory in full action, although the computational capabilities were obviously modest and fully self-consistent total-energy calculations would appear on the scene only a little later. There was emphasis on approximative models and band structures, but the theoretical underpinnings were provided by the density-functional theory of Walter Kohn and his colleagues.

Around the same time, we started a small electronic structure group also at my home base, Helsinki University of Technology. It consisted of Matti Manninen and me (still in Cambridge, but we collaborated through long letters – no e-mail those days), and an occasional younger student. It was great fun indeed. We came up with new ways of solving numerically efficiently all kinds of jellium problems, an activity that I have intermittently exercised to this day.

As a post-doc, I went to NORDITA in Copenhagen. Again, I was very lucky with my timing. John Wilkins was at NORDITA then, and he brought in a large number of visitors sharing an interest in electronic-structure theory and density-functional methods. Obviously, the proponents of the Swedish Electron Gas were closeby in Gothenburg and Lund. It was most productive and enjoyable, and again I made lots of good friends with many of the future key players – including John Perdew, David Langreth, Ove Jepsen, René Monnier and many many more, among them Bengt Lundqvist and Jens Nørskov in Aarhus.

My first faculty position was in the University of Jyväskylä, a small school in central Finland. The research in the physics department was at that time almost exclusively in low-energy nuclear physics. I was the only person in condensed-matter physics, and I also had to cater for "applied physics" courses such as electronics and instrumentation (I do not think I did this very well). However, I managed to keep my electronic-structure activity alive, in contact with my friends at Helsinki University of Technology. Eventually condensed-matter physics took root in Jyväskylä, where there now is a blossoming activity, with Matti Manninen very much in charge.

I spent a sabbatical year at Cornell University, working with Neil Ashcroft. During my stay Joe Oliva came as a fresh post-doc from La Jolla. He had been a Ph.D. student with Walter Kohn, and his thesis dealt with the question of moderated, low-energy positron beams interacting with metals and semiconductors. Such beams had recently been prepared at Bell Labs, and I had also taken an interest in them. Joe's thesis contained many nice ideas, obviously influenced by his supervisor, which we went on to develop and publish together in a paper that was well received.

Back at Jyväskylä, I started thinking about generalising density-functional theory to two-component, electron-positron systems. Eventually, Edward Boronski and I did this and developed a method to calculate the electron and positron densities in solids in a true self-consistent fashion. Actually, it was not too difficult: all we had to do was to follow in the footsteps of the original density-functional theory papers and to try avoid making mistakes.

Another sabbatical at Cornell followed. This time John Wilkins had gathered a group of people planning to have a fresh look at the art of doing electronic-structure calculations, including Ken Wilson, Priya Vashishta, Cyrus Umrigar, John Rehr and Steve White. The landmark paper by Roberto Car and Michele Parrinello had come out a little earlier, and inspired also us to experiment with novel ideas and approaches. I do not think we actually accomplished much during that year, but again the exposure to new viewpoints and angles of attack was to prove very useful later.

Towards the end of the 80's I returned to Helsinki. We started building up the electronic-structure group there, Martti Puska now playing an important role. At the same time, we started promoting the general idea of large-scale computing in a serious way in Finland (inspired by Ken Wilson's success at Cornell). A national high-performance computing center was started in the campus, and I spent quite a lot of my time in trying to stabilize it. The community of physicists and chemists engaged in computational studies had grown substantially and in fact became the largest user group of the new facilities. The Cray machines were happily crunching Kohn–Sham equations day and night ... and do so even today.

We became very much involved in large-scale electronic structure calculations. The group grew and started also many other activities in condensed-matter and materials theory, but in the hard core of the activities density-functional theory rules. It has all kinds of extensions: time-dependent systems opening the realm of dynamical response and transient phenomena; current-density-functional methods enabling attacks on systems at large magnetic fields; multi-component systems etc. Many generations of graduate students have studied and enjoyed the delightfully clear papers by Walter Kohn and his colleagues. My experience is that by far the most popular sections in the introductions of Ph.D. theses are those memorable chapters in the classic Hohenberg–Kohn–Sham papers from the mid 60's ... .

In 1999, I went to Gothenburg to a symposium to honor Walter Kohn and his impact in condensed-matter sciences. It was organized by Bengt Lundqvist and attended by a large number people from the Nordic countries. Professor Walter Kohn had been awarded the Nobel Prize in chemistry the previous year. The symposium was a memorable event, highlighted of course by the personality of Walter Kohn himself. It seems to me that the impact of his work has been especially strong in Sweden, Denmark and Finland. Of course, there is a wide and active European scene, symbolized by the famous $\Psi_k$ community, but in relative terms the Nordic area has embraced perhaps warmest his approach to electronic-structure problems. This was also felt at the symposium. Later, Walter Kohn came to Finland and delivered his lecture to a hall packed to capacity. The event was recorded and remains as one of the most popular videos-on-demand in the archives of the Center for Scientific Computing in Finland.

Density-functional theories (and the other major achievements of Walter Kohn) thrive today more than ever. They are at the heart of the atomistic length-scale of computational materials science, and also among the key tools in nanosciences so fashionable today. They cross the old barriers between condensed-matter physics, chemistry and biology. I consider myself very fortunate for having been able to follow their triumph for more than three decades. I also have endless admiration for Professor Walter Kohn for his personal courage and his uncompromising defense of human values. I know for a fact that there are scores of colleagues throughout Finland who want to join me in wishing him well on his $80^{th}$ birthday. And we all wait for the next paper.

**About the Author**: Risto Nieminen is Academy Professor at Helsinki University of Technology and senior scientific adviser at the Center for Scientific Computing in Espoo, Finland. He is the director of COMP, a national Center of Excellence nominated by the Academy of Finland, specialising in theoretical and computational condensed-matter and materials physics. His research covers electronic, optical and structural properties of materials, including semiconductors, magnetic compounds, materials processing, and quantum nanostructures. His current interests include developing methods for multiscale modelling, which connects the atomistic length and time scales to meso- and macroscales. Contact: Laboratory of Physics, Helsinki University of Technology, P.O. Box 1100, FIN-02015 HUT, Finland; rni@fyslab.hut.fi; www.fyslab.hut.fi

# Walter's Group Parties

Qian Niu

University of Texas, Austin, U.S.A.

During the three years between 1987 and 1990 when I was a postdoc of Walter at UCSB, I had the opportunity of having numerous one-on-one discussions with him and listening regularly to his slow, precise and insightful comments during our group's Wednesday lunch meetings. I also enjoyed greatly the group parties that Walter held at his home every few months then. In the typical air-conditioned Santa Barbara sunny days, the dinner tables were usually placed on a patio in the sloping backyard, with the spectacular wide view of the Pacific Ocean in the distance and Mara's colorful and neatly kept garden in the immediate surroundings. Walter warmly greeted our families, served delicious barbecue cooked by himself, and told us the fine, fine stories. These were the most beautiful moments of life.

I visited Walter a few times afterwards. In the Fall of 1998, I participated in an ITP workshop on the quantum Hall effect. Walter invited me again to his home, together with several of his young associates including Ann Mattsson. Before letting us start to eat his deliciously cooked barbecue, he asked us to tell a most happy event occurred to each of us in the past year. Ann said that her paper with Walter on edge electron gas had just been accepted by PRL. I said that I had received a federal research grant after a few years of struggle. So it went with each guest, and we cheered with drinks to our achievements. Walter then announced the start of the dinner. I reminded him, "Walter, we also want to hear your own story." He thought for a moment and said, "Well, as Ann told you, our paper has been accepted by PRL." I objected, "It doesn't count. She has already said that." So he thought for another moment, but did not come up with something else, ending up saying "give me a break" with a laugh.

I regretted afterwards, feeling guilty for having possibly created a moment of embarrassment. However, all this feeling was completely swept away a week later, when we all heard the news that Walter won a Nobel Prize!

Walter cooking barbecue at a group party in his backyard (1989)

**About the Author**: Qian Niu is a Trull Centennial Professor of Physics at the University of Texas at Austin, a fellow of the American Physical Society, and an editorial board member of the International Journal of Modern Physics B and Modern Physics Letters B. His research concerns quantum transport of electrons in solid state systems and of gas atoms at ultra-cold temperatures. His current interest include Berry phase effects in condensed matter, anomalous Hall effect, and quantum dynamics of Bose-Einstein condensates. He holds a patent on a single electron device, and is a co-author of the book The Geometric Phase in Quantum Systems (Springer, in press). Contact: Department of Physics, University of Texas, Austin, TX 78712, U.S.A.; niu@physics.utexas.edu; niu2.ph.utexas.edu

# He Did Effective-Mass Theory Too!

Sokrates T. Pantelides

Vanderbilt University, Nashville, U.S.A.

Walter Kohn will of course be remembered for ever for Density Functional Theory. His other contributions to physics are also well known. I will mention here the one that affected my years as graduate student at the University of Illinois (1969–73). In some sense, this is how I first "met" Walter. I was working with effective-mass theory for hydrogenic impurities in semiconductors and dug up its history. In the old engineering library I found a two-page document from 1942 stamped in red ink "TOP SECRET", written by Hans Bethe, who articulated the basic idea of "donator" and acceptor impurities in semiconductors. Indeed, the potential of doping semiconductors was appreciated at a crucial time and led to the transistor in 1948. There were a few other early formulations of effective-mass theory. Then came Walter's papers in the 1950's. They offered a very thorough, rigorous, and detailed derivation with all the approximations spelled out, accompanied by careful and deep, insightful analysis. For me, those papers were a first lesson in theoretical physics. Effective-mass theory was one of the items for which Walter was awarded the APS Buckley Prize in 1961 (before DFT!).

I met the real Walter Kohn for the first time when I was a post-doc at Stanford, 1973–75. I believe it was at a March meeting at some reception. I told him about my thesis work using effective-mass theory and I was pleasantly surprised by the genuine interest he showed in the way I used the theory, what I got out of it, and the method's potential for additional work. At that time, DFT was still relatively dormant. Within years, it was established as the method of choice, especially when total-energy calculations became absolutely necessary, first with surface reconstruction and, soon after, the area I embraced in the late 1970's, point defects. Walter was my hero again, everybody's hero.

I ran into Walter on many occasions since my initial encounter and only have fond memories. The latest was when he visited Vanderbilt for a Distinguished Lecture with his wife. It was truly a memorable visit as both Walter and Mara gave lectures that were attended by good crowds. It was great fun talking physics on the blackboard ... .

**About the Author**: Sokrates T. Pantelides is the William A. and Nancy F. McMinn Professor of Physics at Vanderbilt University in Nashville, TN. He holds a concurrent appointment as Distinguished Guest Scientist at nearby Oak Ridge National Laboratory. Prior to joining Vanderbilt in 1994, he spent 20 years at the IBM T. J. Watson Research Center in Yorktown Heights, NY. His research employs Density Functional Theory to probe atomic and electronic properties of solids. Recent work has been on complex systems such as the Si-SiO$_2$ interfaces, polycrystalline electronic ceramics such as SrTiO$_3$, catalysis using complex materials such alumina and nanocrystals, and electronic transport in molecules and other nanostructures. Contact: Vanderbilt University, Department of Physics & Astronomy, Nashville, TN 37235, U.S.A.; pantelides@vanderbilt.edu

# Thoughts and Recollections of a Remarkable Man

Dimitrios A. Papaconstantopoulos

Naval Research Laboratory, Washington, U.S.A.

While I had occasionally met Walter Kohn over the years at APS and other meetings, it was in 1997, at a meeting on electron correlations in Crete, that I became aware of some sides of Walter other than those associated with his scientific accomplishments. At the meeting, I walked with him on the beach and had lunch and dinner a few times. We talked about many things other than physics, and I enjoyed the fact that he was very interested in Greece, obviously a favorite topic of mine. What impressed me about Walter was his interest in everything and everyone around him, as well as a complete absence of the ego one often encounters in famous scientists. He was comfortable in every situation except, amusingly, when everyone was throwing and breaking plates during a traditional Greek dance. I couldn't get him to toss even one plate.

At that meeting I expressed, in his absence, the prediction and hope that he would receive the Nobel Prize for his work on Density Functional Theory. The following year, in fact, he won the prize and I felt, like many others, delighted and vindicated because of the many years spent using his theory. We met at a few more meetings, one of them in Vienna in January 2001 to honor Karlheinz Schwarz's 60$^{th}$ birthday. We walked together every morning from the hotel to the Technical University. While we walked I listened, fascinated, while he described his remarkable life, his early years in Vienna, the war, and his politics.

In February 2001, I invited Walter to NRL to give a talk and offer advice on the work of my group. He spent two days with us. First he met with a couple of administrators at NRL where he asked penetrating questions about our organization and offered thoughtful opinions. This is one of the most striking characteristics of Walter Kohn, his ability to be completely engaged by what he is experiencing. As a result of this engagement a wonderful part of the day ensued when I asked him to listen to presentations from ten scientists in my group. I had estimated about one and a half hours for these presentations. In fact, it took us four hours. Walter questioned, praised, and criticized everybody. It was a memorable experience for all of us. My colleagues felt good

Dimitrios A. Papaconstantopoulos

about their research, were flattered by his attention, and benefited from his corrections and insights.

The next day we were treated to a lecture on DFT that was a fascinating mixture of history and eloquent exposition of the theory. When a renowned scientist visits, it is almost always a productive experience. But the short time Walter spent with my group left us with much more, a renewed excitement about our work.

**About the Author**: Dimitrios A. Papaconstantopoulos is Head of the Center for Computational Materials Science at the Naval Research Laboratory in Washington and Professor of Computational Science at George Mason University. His areas of expertise include band structure methods, superconductivity, theory of alloys, and construction of tight-binding Hamiltonians. Contact: Center for Computational Materials Science, NRL, Washington, D.C. 20375-5345, U.S.A.; papacon@dave.nrl.navy.mil; www.nrl.navy.mil

# The Bonding of Quantum Physics with Quantum Chemistry

Robert G. Parr

University of North Carolina, Chapel Hill, U.S.A.

The bond that developed between quantum physics and quantum chemistry, that led to the award of a big chemistry prize to the physicist Walter Kohn in 1998, developed not without trial. Here I give an account of it. An element in this bond has been a friendship between Walter Kohn and me. My having reached 80 first, he has already kindly spoken of this[1]. Now it is my turn.

In the 20s and early 30s there was a flush of successes in establishing the ability of quantum mechanics to describe the simplest molecules accurately: the Born-Oppenheimer approximation, the nature of chemical bonding, and the fundamentals of molecular spectroscopy. But then the quantitative theory of molecular structure, which we call quantum chemistry, was stymied, by the difficulty of solving the Schrödinger equation for molecules. The senior chemical physicists of the 30s pronounced the problem unsolvable. But the younger theoreticians in the period coming out of WWII thought otherwise. Clearly one could make substantial progress toward the goal of complete solution, because the equation to solve was known and had a simple universal structure. The boundary conditions too were known. It would not be as easy as handling an infinite periodic solid, but a number of us set to work. The special demand of chemistry was to quantify very small molecular changes. Successes came slowly, but with the development of computers and a lot of careful, clever work, by the 90s the quantitative problem was essentially solved. The emergent hero of the chemical community was John Pople, whose systematic strategy and timely method developments were decisive. The methods of what is termed "ab initio" quantum chemistry became available and used everywhere.

Over the years the quantum chemists did a lot more than gradually improve their ability to calculate wavefunctions and energies from Schrödinger's equation. All the while they have served molecular spectroscopy, physical inorganic chemistry, and physical organic chemistry. Relevant for the present story was the development by Per-Olov Löwdin in 1955 of the density matrix

---

[1] W. Kohn. In: Reviews of Modern Quantum Chemistry. (Ed.) K.D. Sen. Vol. I, II, World Scientific, Singapore 2002, pp. v-vii

reduction of the Schrödinger equation, especially the identification and mathematical physics of natural spin orbitals and their occupation numbers. The hope was, although hope floundered, that the Schrödinger problem could be resolved in terms of the first- and second-order density matrices. Foundering came because of the difficulty of incorporating the Pauli principle.

Beginning way back in the 20s, Thomas and Fermi had put forward a theory using just the diagonal element of the first-order density matrix, the electron density itself. This so-called statistical theory totally failed for chemistry because it could not account for the existence of molecules. Nevertheless, in 1968, after years of doing wonders with various free-electron-like descriptions of molecular electron distributions, the physicist John Platt wrote[2] "We must find an equation for, or a way of computing directly, total electron density." [This was very soon after Hohenberg and Kohn, but Platt certainly was not aware of HK; by that time he had left physics.]

From the end of the 40s, I was a happy participant in most of these things, ab initio and the rest, although from about 1972 I became pretty much an observer. We plunged into density-functional theory.

DFT soon intoxicated me. There were the magnificent Hohenberg–Kohn and Kohn–Sham 1964–65 papers. The Xalpha method of John Slater was popular in those days, but it was not sufficient for the high accuracy needed. And I was much taken by the work of Walter Kohn, whom I had known since 1951. There were many things to do: Improve upon the LDA to reach the accuracy needed for chemical applications. Shift the emphasis on fixed, very large electron number toward variable, small number, since that most concerns chemistry. Enlarge the language to include chemical as well as solid-state concepts. Introduce into DFT, as appropriate, some of the theoretical advances already made within quantum chemistry. All of these things subsequently came about. The methods and concepts of DFT became available and used throughout the chemical community.

I had been on the faculty of Carnegie Institute of Technology for a couple of years when Walter Kohn arrived in 1951. I was aware from the beginning of the strength of physics at Carnegie, especially solid state physics. Fred Seitz was the Head when I arrived, and several other solid state experts also were there. I bought Seitz's great book for $ 6.38 and browsed in it, noting in particular the fine description of the Hartree–Fock method (but not finding any treatment of the invariance to unitary tranformation of orbitals that is so important for understanding the equivalence of localized and non-localized descriptions of molecules). I enjoyed pleasant interactions with a number of the physicists. Soon after Kohn arrived I had two physicist postdocs, Tadashi Arai from Japan and Fausto Fumi from Italy, who became acquainted with him. On the thesis examination committee of Walter's graduate student Sy Vosko, I learned that it was okay to use trial wavefunctions with discontinuous

---

[2] John R. Platt, letter to RGP, dated October 23, 1968

Pierre Hohenberg, Walter Kohn, and Robert Parr in Stockholm, December 1999 (photo: P. Hohenberg)

derivatives. I was pleased to attend an evening party at the Kohns, and I was disappointed when Walter left Carnegie for elsewhere.

I do not recall when I first heard of the Hohenberg–Kohn–Sham papers, but I do know that the quantum chemistry community at first paid little attention to them. In June of 1966 Lu Sham spoke about DFT at a Gordon Conference. But in those days there was more discussion about another prescripion that had been on the scene since 1951, the Slater Xa method. The Xa method was a well-defined, substantial improvement over the Thomas–Fermi method, a sensible approximation to exact Kohn-Sham. Debate over Xa went on for a number of years. Slater may never have recognized DFT as the major contribution to physics that it was. [When I asked John Connolly five or six years ago how he thought Slater had viewed DFT, he replied that he felt that Slater regarded it as "obvious".]

Walter Kohn's appearance at the Boulder Theoretical Chemistry Conference of 1975 was memorable. On June 24 he presented a formal talk, in which he outlined DFT to the assemblage of skeptical chemists. There were many sharp questions and a shortage of time, so the chair of the conference decided to schedule a special session for the afternoon of June 26. With quite a crowd for an informal extra session like this, Walter held forth on his proof. In his hand he held a reprint of the HK paper, from which he quietly read as he slowly proceeded: " ... and now we say ... ". The audience sobered down quickly. It was a triumph. The interest of quantum chemists in DFT began to grow at about this time.

Our group began contributing to DFT in the 1970s. In some of our first work, my graduate student Robert Donnelly generalized the original idea to functionals of the first-order density matrix. In 1977 I described the central result at Walter Kohn's luncheon seminar in San Diego: All natural orbitals with non-zero occupation numbers have the same chemical potential. Discussing this with Walter at the blackboard afterwards, I remember his saying "This must be correct". [Walter himself recently recalled this incident.] First-order density matrix functional theory is receiving fresh attention nowadays.

As we ourselves kept plugging along, the quantum chemical community largely was negative about DFT, even antagonistic. Their "house journal" International Journal of Quantum Chemistry, in 1980 published a pointed criticism of it[3]: "There seems to be a misguided belief that a one-particle density can determine the exact N-body ground state". In 1982 Mel Levy and John Perdew replied with a letter that was both incisive and eloquent[4]: "The belief is definitely not misquided". Yet, in the same issue of IJQC, the editors called for further discussion of the "controversial" subject[5]. It was going to be awhile before quantum chemists were convinced.

Over the period 1979–1982, Mel Levy supplied a major advance with introduction and careful discussion of the constrained search formulation of density functional theory . This greatly heightened confidence in the theory (and it still does!). Then in 1983 came Elliott Lieb's masterly detailed analysis, which validated DFT as rigorous mathematical physics. [Once in the 70s I asked Barry Simon, the mathematician who with Lieb had done a famous rigorous analysis of the Thomas-Fermi theory, what his opinion was of DFT. "It may be good physics", he said, "but it is not good mathematics". Lieb's paper signaled the end of the period of doubt about DFT. The space for further development was now wide open and the interest of chemists began to accelerate.

What computational chemists wanted above all else was calculational methods for molecules, and the LDA just was not enough. The need for more accurate exchange-correlation functionals was met in the 80s, with an accuracy that has proved quite good enough for the times. The Nobel award in 1998, one may point out, was specifically designated to be a prize for computational chemistry. Well, good, and immensely deserved in my opinion. I note, however, that there is another whole side of DFT which has concerned and still concerns many of us, the "conceptual" side. This side is rich in potential, and it is not without accomplishment. The concepts of DFT neatly tie into older chemical reasoning, and they are useful for discussing molecules in course of reaction as well as for molecules in isolation. Where solid-state physics has Fermi energy, chemical potential, band gap, density of states, and local density of states, quantum chemistry has ionization potential, electron

---

[3] International Journal of Quantum Chemistry **18**, 1029–1035 (1980)

[4] M. Levy and J.P. Perdew, International Journal of Quantum Chemistry **21**, 511 (1982)

[5] International Journal of Quantum Chemistry **21**, 357 (1982)

Robert Parr and Walter Kohn with a Geisha in Kyoto, mid 1990s

affinity, hardness, softness, and local softness. Much more too. DFT is a single language that covers atoms, molecules, clusters, surfaces, and solids.

Walter Kohn has been a great help to many scientists over many years, an expert consultant and helpmate and a fine, unobtrusive, even-handed host of good meetings in lovely places. We thank him. In recent years I have discussed with him (among other things), circulant orbitals, the monotonic density theorem, and the information theory point of view on what constitutes an atom in a molecule), the latter during a stolen few minutes in a Stockholm hotel in December of 1999[6]. Walter may or may not "like" chemistry[7], and he claims not to have studied chemistry in the university. But what does one call a great teacher of chemical principles? I would say, CHEMIST, full caps.

**About the Author**: Robert Parr is Distinguished Professor in the Department of Chemistry, University of North Carolina. Author (with Weitao Yang) of Density-Functional Theory of Atoms and Molecules (Oxford), he has been doing research in quantum chemistry continuously since the late 1940s. Contact: Department of Chemistry, University of North Carolina, Chapel Hill, N.C. 27599-3290, U.S.A.; rgparr@email.unc.edu

---

[6] RGP was delighted to be Walter Kohn's guest. See photograph on page 189 and P. Hohenberg, this volume, p. 100

[7] E. Eliel, This volume, p. 66

# My Meetings with Walter Kohn

Michele Parrinello

ETH Zurich and Swiss Center for Scientific Computing, Manno, Switzerland

I have a very clear memory of the first time I saw Walter. It was in the early seventies, at a conference on polaritons at Taormina in Sicily, and it was a bright, sunny day. At the time I was a young post-doc at the University of Messina, Walter was already very famous and I and my young colleagues regarded him with admiration and awe. For this reason, but also because of the linguistic barrier, we simply did not dare to approach him.

Our first real meeting took place many years later in Trieste, where he was a regular summer visitor. Characteristically, he was going round talking to people in order to find out what they were doing, and he came to speak to me about the newly-developed Car-Parrinello method. Once again I have a distinct memory of that meeting and his probing questions, while sitting on the terrace of the Adriatico building of ICTP.

From then on, our meetings became more frequent and our friendship began to flourish. I remember flying to Paris just to discuss with him the problem of O(N) algorithms and above all an evening in Zurich, another place he visits regularly, when he kept us spellbound with tales of his youth in Vienna, the sad moment when he had to leave that city, the life of Jews in eastern Europe and the work of Elias Canetti. Walter is a riveting raconteur and this is a memory I will cherish for ever.

In another, more recent meeting, he amused me with his description of the Nobel prize-giving ceremony and how by mistake he was served a Kosher meal, so that while the rest of the crowd was feasting on a wonderful spread served on exquisite porcelain, he had to make do with a piece of boiled chicken on a plain white plate.

I consider myself lucky to have enjoyed Walter's friendship and have to confess one special, personal fact that makes me like him even more. He looks just like my grandfather Michele.

**About the Author**: Michele Parrinello is Director of the Swiss Center for Scientific Computing (CSCS) and Professor at ETH Zurich. Together with Roberto Car he introduced the ab-initio molecular dynamics method, which he is still developing and applying. His scientific interests include the study of complex chemical reactions, hydrogen-bonded systems, catalysis and materials science, in particular systems under pressure. He is also known for the Parrinello–Rahman method of molecular dynamics, which allows the study of crystalline phase transitions under constant pressure. Contact: Swiss Center for Scientific Computing (CSCS), Via Cantonale, CH-6928 Manno, Switzerland; parrinello@cscs.ch; www.cscs.ch

# The Greatest Gift Is Something Worth Thinking About

John P. Perdew

Tulane University, New Orleans, U.S.A.

Walter, I was never your student, postdoc, collaborator, or co-worker. Yet most of my career in science since graduate school has been shaped by your intellectual and personal influence. In the density functional theory (DFT), you gave us something worth thinking about. That is a great gift, one of the greatest, one that has provided me with many years of perplexity and enjoyment.

As a graduate student at Cornell, and in my first year as a postdoc with Sy Vosko at Toronto in 1971, I thought about electronic structure theory in terms of pseudopotentials screened in the Hartree approximation. Then Mark Rasolt came down from Ottawa as Sy's second postdoc, bringing the word that there was a density functional theory (news to me but not to Sy), and that it worked like a charm (news to both of us, I suspect). Sy and I immediately got interested and stayed interested in this theory. In my second postdoc appointment with David Langreth at Rutgers, the hidden reasons for the success of the local density approximation started to emerge, along with the idea that the second-order gradient expansion could be corrected – the beginning of the generalized gradient approximations now widely used. From David, I learned how to make things right by taking away what is wrong.

After I started my faculty position at Tulane University in 1977, I learned that your good influence had again preceded me. A panel had recommended that Tulane's doctoral program in physics be discontinued. The University then invited you in as a consultant in 1976, and you recommended instead that the program be built up. And I've enjoyed working here ever since.

In 1982, with Mel Levy and Bob Parr, I discovered something fascinating about your exact density functional: Since separated atoms have integer electron numbers, the functional must have a discontinuous derivative at the integers. It was exciting to discuss this at the 1983 Institute for Theoretical Physics workshop on Many-Body Phenomena at Surfaces. Since then, we've met at innumerable conferences in many an exotic locale.

In 1989, I was asked to nominate for the Nobel Prize in Physics. I nominated you, as many have done before and since. Although my nomination was unsuccessful, I have the consolation of having been ahead of my time (or more precisely the Nobel Committee was behind the times). When you did win the Prize in Chemistry in 1998, there were celebrations here at Tulane. Not bad for someone who admits to taking his last chemistry course at the age of 16!

When I interviewed you a year ago for the Vega Science Trust (http://www.vega.org.uk), I asked you what advice you would give to young scientists. You recalled Eugene Wigner's advice to scientists: "When you get a result, try to get it another way." That's characteristic of you – never satisfied with the first answer, and always digging deeper. At an age when many are content to rest, you are still a dynamo of thought and action.

I hope you will enjoy this book written by the Kohnoscenti. Happy birthday, and thanks for the gift.

**About the Author**: John P. Perdew is a professor and chair of the Department of Physics at Tulane University. His research is in condensed matter physics and density functional theory. He is interested in finding the properties of the exact density functional for the exchange-correlation energy, and building those properties into approximations that may be useful in both condensed matter physics and quantum chemistry. He has *never* violated the uniform electron gas limit, but is nevertheless one of the 100 most-cited physicists 1981-present (http://www.isihighlycited.com). Contact: Department of Physics, Tulane University, New Orleans, LA 70118, U.S.A.; perdew@tulane.edu

# Vignettes:
# Switzerland, Australia, Santa Barbara

Warren E. Pickett

University of California, Davis, U.S.A.

I must not be alone in how I remember specific locations in connection with vivid memories of Walter Kohn. I can briefly relate strong impressions that Walter has left me with, from three spots across the globe.

It must have been in 1991, that Walter and I happened to be visiting the University of Geneva on the same one or two days. For that reason I found myself in a Kohn seminar with one aspect that renders it almost unique in modern times. Walter was particularly excited at this time about some new experiments that were being done (in the Chicago area, I feel almost sure) where electrons were being extracted from surfaces with extremely large electric fields. There was not much theory yet, no doubt it was this void that intrigued Walter.

What was memorable about this seminar? one might ask, especially since I have forgotten all of the specifics. The striking realization, as he neared the end of his 45–50 minutes, was that he had used *three* vugraphs. (My memory is only good to about 30%; it might have been *four*.) He had not "presented a seminar" in the conventional sense that almost all of us think in terms of: "I've got my 24 vugraphs so I'm ready to give a seminar." He had come to talk to the audience with only three (or four) visual aids. Few scientists would want to attempt such a daring venture.

It was at Brisbane, Australia, during a meeting organized and hosted by John Dobson in 1996, where Walter introduced (to me, and I think the world) the initial results on his studies of electrons at surfaces (which were known for some time to be a problem for the "standard model" that we call LDA). He simplified the system to a collection of electrons in a linearly increasing (ramp) potential, which led approximately to some solutions that mathematicians had known about for a century, probably, and had of course named. Walter's results at that point were preliminary, as I recall; in fact, still today he is pushing this area of study forward. What was particularly memorable was the play on words of the title he gave to this system: the "Airy gas." As opposed to the "steely metals" he had often studied in the past, or perhaps to contrast to "leaded gas"?

The most recent memory that I will relate was at (his) home base, while I was spending four months at ITP at UCSB. As one of the coordinators of the quantum magnetism program that was running at the end of 1999, I was mindful to stimulate interaction wherever I could. Moreover, I had a pet problem that I wanted to see tossed around: does spin density functional theory have something unusual or unexpected to say about half metallic ferromagnets, or perhaps vice versa. I managed without much difficulty to assemble Walter Kohn and two other program attendees, Helmut Eschrig and Igor Mazin, for a very lively discussion of the general question. We all met again about a week later; it was evident that some parts of the discussion were getting even "livelier." Then the unfortunate fact arose that Walter had to leave on what I seem to recall was his annual 'grand tour' of Europe in the fall and early winter. This work resulted in due time with an important publication, but it was a shame that Walter's collaboration and coauthorship was not to be a part of it.

To conclude with memories: it was with pleasure that I was able during this same period to attend the modest celebration at ITP when Nobelist Walter Kohn was able to bring in his Nobel Prize to ITP for the first time. This was in 1999; it was of course in 1998 that Walter was awarded the prize, but Mara's illness at that time (as I learned) was more on Walter's mind than accepting the prize, and the Committee agreed to allow him to accept his prize formally the next year. It was thus in late 1999 that Walter brought his prize periphenalia (some metal, some artwork) to share with friends and colleagues.

**About the Author**: Warren Pickett is professor of physics at the University of California Davis and an editor of the Journal of the Physics and Chemistry of Solids. His research focusses on the microscopic description of solids that display unexpected phenomena (such as competing order parameters) arising from underlying complexity (structure, interactions, etc.) Most of these behaviors involve magnetism and/or superconductivity, and many invite improvements in the underlying theory. Contact: Physics Department, UCD, Davis, CA 95616, U.S.A.; pickett@physics.ucdavis.edu; yclept.ucdavis.edu

# Theory Versus Reality

John J. Rehr

University of Washington, Seattle, U.S.A.

Physical theories and their practical realizations are complementary. Walter Kohn's theories are noteworthy both for their originality and their elegant physical and mathematical underpinnings, and they have had a profound influence in the development of condensed matter physics. Yet the practical realization and acceptance of such theories often takes years. In his book *Disturbing the Universe*, Freeman Dyson recalled asking J. Robert Oppenheimer what direction he should follow in physics and receiving a seemingly cryptic answer:

"You should follow your destiny." What I have come to appreciate is that the realization of Kohn's theories has become the destiny of many of us. I was fortunate to have been a Postdoc in Walter's group at UCSD, La Jolla from 1973–75. At that time UCSD was a Mecca for condensed matter physics, and there were many famous pilgrims. The Nobelist John Bardeen often visited and always made the rounds to find out what each of us was doing. The Russian physicist Alexei Abrikosov once visited, and I had the opportunity to drive him to Disneyland – then a favorite stop of first time visitors from the USSR.

It was a challenge, at least for me, to keep up with Walter whenever we discussed physics. But the clarity of his ideas, expressed with good humor and encouragement, often suggested good starting points and made working with him a pleasure. In Walter's case, Voltaire's subtle "smile of reason" was usually an infectious laugh full of genuine enthusiasm.

For me, the development of a local description of condensed matter, a task that started at La Jolla, became a long term goal. Walter had a strong belief in the principle of locality and it became central to my thinking as well. In contrast to the long range nature of one-particle wave functions in solids, much of the physics in condensed matter is short-ranged. The standard reciprocal space methods often don't work well locally. Yet as Walter anticipated, Wannier functions which describe local electronic structure remain exponentially localized even in aperiodic systems. Scattering theory and its applications became another part of my destiny, inspired by Walter and some of his work on scattering with Res Jost and Julian Schwinger.

A quarter of a century later, many of Walter's theories have become standard techniques. Exponential advances in computer power have enabled physicists to make quantitative calculations based on his ideas.

Certainly Walter's Nobel Prize was due in part to the success of his density functional theory and the local density approximation. But those of us who know Walter will also appreciate him for his inspiration.

**About the Author**: John J. Rehr is a Professor of Physics at University of Washington specializing in condensed matter theory with an emphasis on computational techniques. His major research interests include theories of real space electronic structure, x-ray spectra, and electronic excitations. He is a Co-coordinator of the DOE Computational Materials Science Network (CMSN), and his group is especially known for its X-ray spectroscopy codes. Contact: Dept. of Physics, University of Washington, P.O. Box 351560, Seattle, WA 98195-1560, U.S.A.; jjr@phys.washington.edu

# Recollections of Walter in La Jolla and Zurich

Maurice Rice

ETH Zürich, Zürich, Switzerland

One of the most exciting moments of my life was in the spring of 1964 when I received Walter Kohn's offer of a post-doc position in La Jolla. I was finishing up my Ph.D. but since my Irish scholarship had run out, I had moved from Cambridge to Birmingham to take a temporary teaching position. The combination of a top class mentor and beautiful sunny Southern California, seen from the perspective of the grey English midlands was overwhelming. I had heard only good things from my fellow student and friend Lu Sham who had joined Walter Kohn as a post-doc in the previous fall. So I arrived in La Jolla with high expectations in the fall of '64.

Shortly before I arrived the physics department of the new La Jolla campus had moved up onto the mesa, so I missed the experience of life at the beach in the temporary quarters at the Scripps Institute of Oceanography. But for me it was still magical to work amid eucalyptus trees with a view of the ocean in an atmosphere of great optimism and enthusiasm for the future of solid state physics. There were pessimistic voices at the time (c.f. Brian Pippard's article in Physics Today with the title "The Cat and the Cream" with the clear message the cream was all gone and we should only look forward to a diet of skimmed milk). But at La Jolla there were great hopes for this new campus and its stars recruited from all across America. I remember Walter talking at lunch about his vision of a universal scheme to calculate the properties of the electron fluid which could be applied to metals, semiconductors etc. Of course he and Lu Sham were working together to realize this vision and they succeeded in a way that I believe surpassed even Walter's hopes.

To me Walter suggested a different topic. At that time, the mid-sixties, there was a lot of interest in superconductivity in low dimensions, possibly as a way to reach higher transition temperatures as Bill Little at Stanford was proposing. However Walter's formidable physics intuition told him that fluctuations would become very important in low dimensions. This was before the great progress on critical phenomena and the role of fluctuations that came a few years later. I didn't know much about superconductivity or the role of fluctuations but the topic piqued my interest. Walter had an organized life and scheduled a weekly discussion hour for me with him. As it happens this was on Monday morning, which was a mixed blessing. As each week would come to an end, I would be faced with the imminent problem of what progress would I be able to report on the following Monday. This scheme is one that encourages post-docs to work on the weekends and held my natural inclination towards the life of a beach bum in check.

This weekly attack of anxiety especially occurred at the beginning of the project when we were unclear how to formulate the problem. Around this time John Bardeen came to La Jolla for a longer visit. The La Jolla campus was of course very attractive to mid-westerners in Winter. Perversely the local TV stations loved to show pictures of the less fortunate mid-westerners digging out their cars from snowstorms in winter so as to convince us how right we were to have chosen to live in Southern California. Naturally Walter scheduled a discussion for both of us with John Bardeen on this topic and naturally as a young green post-doc, I was quite intimidated to be discussing with such eminent theorists. John Bardeen was a man of few words and given to Delphic utterances. At that time Walter Kohn liked to smoke a pipe. Now Walter does not like to make rash statements and prefers to consider his remarks carefully. When a particular knotty point would come, I remember Walter would suddenly become preoccupied with his pipe, knocking out the ashes, refilling it with new tobacco etc. I must confess to the suspicion that it wasn't purely a coincidence that he would start working on his pipe when the question under discussion was particularly subtle and complex. However I heard from the other students and post-docs that the worst thing that one could do was to interrupt these pipe pauses with banal and ill thought out comments. Rather one should wait until Walter was ready to give his answer. Coming back to the discussion between John Bardeen and Walter Kohn, it was marked by long pauses, during which Walter worked a lot on his pipe, interspersed with Delphic remarks from John Bardeen. As a young post-doc I was too intimidated to demand fuller explanations, and mostly sat quietly trying to unravel what John Bardeen was telling us and trying to make an occasional, at least halfway, sensible remark. Eventually John Bardeen announced he had to leave. As soon as the door was shut Walter turned to me and said, "What did John Bardeen say?" Naturally I was relieved to hear that I was not alone in having difficulties following the great man. In due course Walter and I made some progress on the role of thermal fluctuations in destroying off diagonal long range order in two dimensions but when it came time to write up Walter, to my great surprise, decided not to be a coauthor although he had certainly contributed a lot to the project.

The second project that I worked on with Walter introduced me to the subject of metal-insulator transitions – a topic that has stayed with me ever since. Walter had met Jacques des Cloizeaux in Paris shortly before and was intrigued with his observation that a continuous band crossing in an indirect semiconductor should be preceded by the exciton going soft. Therefore at such a metal-insulator transition there should be a phase with the character of a Bose condensed exciton condensate. However as we started to construct a theory for this excitonic insulator state, we found that Leonid Keldysh and Yu. Kopaev had developed a mean field theory for this state which derived the spin density wave state from the semimetallic side as a BCS-like pairing of an electron and hole in the semimetal. Together with Denis Jerome who was spending a year at La Jolla doing theory after his Ph.D., we developed the idea further. The combination of elegant physics and a catchy name has made the excitonic insulator a topic that constantly reappeared in the intervening years. Actually a few years later when I was at Bell Labs, Bill Brinkman and I worked on the electron-hole liquid state in optically pumped Ge and Si again following Leonid Keldysh, and found that the stability of the metallic state of the electrons and holes over the excitonic state generally implied that the metal-insulator transition should be a first order transition from semimetal to semiconductor bypassing the excitonic insulator phase altogether. Last year Walter visited ETH Zurich and I was very happy to discuss with him how Steve Girvin and Allan MacDonald had interpreted the coupled phase of a quantum Hall bilayer with half-filling of the lowest Landau level, very nicely as a form of excitonic insulator. It all goes to show that if the underlying physics is correct, then someday a system will be found that realizes it. Indeed, the whole field of excitonic condensates is currently undergoing an exciting revival.

A visit to the sun on the Rigi mountain to escape the fog that covered the lowlands, in 1986, (left to right, Walter, John Ward, Hilde and Res Jost; photo: courtesy of H. Jost)

Since moving to Zurich some twenty years ago now, it has been a great pleasure and of course a great stimulus to condensed matter physics here, to have had Walter and Mara for many extended visits. They have many friends here. The Josts, Res and Hilde, have been close friends with Walter for over fifty years. Hilde Jost told me they first met at the very first Les Houches Summer School in 1951. Back then it was a very rustic affair located in an old mountain restaurant and at night the participants could hear the mice scampering around in the walls. Walter had come down to the school from Copenhagen where he was a post-doc at the Bohr Institute. A close friendship grew up between Walter and the Josts, partly because of the similarities in their backgrounds. For example, Hilde, like Walter, grew up in Vienna and so Walter could converse with her in their common *wienerische* dialect. On the physics side Walter and Res had common interests in scattering theory. These developed into a fruitful collaboration which was continued with extended visits by Walter to Princeton, where the Josts were based at that time, and by the Josts to Copenhagen the following summer. Hilde remembers that it was a cold Danish summer which was particularly bothersome for the Kohns who had a new born baby, Ingrid. Afterwards they got together regularly. Hilde remembers a visit with the Kohns in New Jersey where Walter was spending a summer at Bell Labs. The Josts drove up from Princeton in weather that is typical for a New Jersey summer, very hot and humid. Their car wasn't air conditioned so they arrived hot and sticky. To Hilde's chagrin Walter proposed a typical European picnic at a park some distance away. When they got there, the picnic ground was deserted as all the local New Jerseyites were avoiding the mid-day heat and lunching in their air conditioned homes. However to Hilde's relief they returned soon to the cool house as the picnic site was very primitive and unsuitable. Another of their meetings Hilde remembers took place in Geneva where Walter's uncle, who lived nearby in Lyon, came to a visit. Hilde remembers him as charming but with a certain French disdain for the highly developed orderliness of Switzerland.

Walter with Res Jost, jr. (1988; photo: courtesy of H. Jost)

Res Jost passed away in 1990 at a time when Walter was visiting ETH. Walter paid tribute to his close friend with a moving eulogy at the funeral service. Walter's relationship with ETH continued to grow and to my and my colleagues' delight ETH conferred an honorary doctorate on him in 1995.

Walter attends the Ascona Conference on Strongly Correlated Electron Systems on the occasion of my 60th birthday in 1999 (photo: G. Blatter)

The Kohn house (photo: courtesy of H. Jost)

Hilde, together with her son Res and daughter-in-law Christine, visited Walter and Mara in Santa Barbara in 2001 (photo: courtesy of H. Jost)

I think also for Walter and Mara Zurich is attractive as a place to revisit central Europe where they spent their childhood. One particular incident stands out in my mind. Walter has always arranged a party as the visits came to an end. At one of these parties he showed me some postcards from Vienna between the wars. He had found these that morning at the Saturday flea market in downtown Zurich. The postcards were published by his father's firm, Gebrüder Kohn, in Vienna. I am sure they brought back bittersweet memories to him. For me it made the terrible tragedy in his early life real in a moving way.

Walter, your many friends and colleagues in Zurich wish you well for your 80th birthday and the years ahead and we will be delighted to see you and Mara in Zurich as often as you can spare the time.

**About the Author**: Maurice Rice is a native of Ireland and studied in Dublin and Cambridge before he started his post-doc with Walter Kohn at UCSD in 1964. Two years later he joined Bell Labs and in 1981 took up his present position in the Institute for Theoretical Physics, ETH Zürich. In recent years he has mostly been preoccupied with the physics of oxides, particularly the high temperature cuprate superconductors. Contact: Theoretische Physik, ETH Zürich, CH-8093 Zürich, Switzerland; rice@itp.phys.ethz.ch

# I am Happy that the R Stands for Rostoker

Norman Rostoker

University of California, Irvine, U.S.A.

In 1950, I completed the requirements for the D.Sc. degree at Carnegie Institute of Technology in Pittsburgh. Walter Kohn arrived in the summer as a new Assistant Professor. We both were undergraduates at University of Toronto. I did not know him there but I worked in the Department of Applied Mathematics for a master's degree where he had finished several years previously. He had a giant reputation for a student.

At Carnegie Tech, Walter had some ideas on how to calculate energy bands in periodic lattices that interested me, so I began to work for him as a "post doc". I had a job as an experimental physicist at Carnegie Tech, so I didn't need to be paid. I worked on the problem for three years – part-time. It was also part-time for him because he went to Copenhagen on leave at the Bohr Institute for more than a year. There were extensive numerical calculations to be done. I provided the resources for this. In those days a computing machine consisted of a noisy electro-mechanical machine called a Marchant. However, the operator of the machine was a cute young girl who was very efficient. Her name was Alice Carroll. Modern methods are undoubtedly much faster but probably Walter would agree with me that there are other considerations; if you enjoy your work, why rush. We did produce some papers on the use of Green's functions for calculating energy bands in periodic lattices. Then we discovered that Korringa had developed essentially the same method with a scattering matrix approach. It is known as the K.K.R. method. I am happy to say that the R stands for Rostoker and the paper is still cited. My contributions were mainly the numerical analysis and helping Walter buy a cheap second hand car that brought him to work without breaking down. I still believe that justifies my being a co-author.

In 1956, I moved from Chicago to San Diego to work at General Atomics. A few years later, UC-San Diego started as a graduate school of Applied Science. They recruited a very high level of staff for the Physics Department, one of which was Walter Kohn. I dreamed of becoming a Professor at this new school. At General Atomics I collaborated with Marshall Rosenbluth and he recommended me for a Professorship. Then I interviewed with the Chairman of the Department of Physics, who was Walter Kohn. At that time, I had back trouble and it was painful to sit in a chair – so I lay down on the floor. It was an extensive interview, during which time many other faculty members came into his office on other matters. They usually noticed me lying on the floor before they left. Walter managed a "dead-pan" response to their surprised looks as though all candidates for faculty positions were interviewed this way.

A year after I joined the faculty, I met Professor Leopold Infeld who knew both Walter and I from University of Toronto. He had moved back to Poland, but was attending a Relativity Conference at University of Texas. For one year when I worked on a M.A. degree, I took all of his courses in spite of the fact that I was a graduate of the School of Engineering. He always called me the engineer in the class and referred to engineers with some disdain. I told him I had recently joined the Physics Department at UC-San Diego, to which he replied "you must be pretty good for an engineer or Walter would not have approved because he is tough!"

During my five years at UC-San Diego I noticed this. There were wonderful moments between Walter and Bernd Matthias, who was the king of superconductivity. Bernd had some disdain for theory. For him, physics truth could only be found in the laboratory. Nobody gave an inch.

After five years I moved to Cornell University as IBM Professor of Engineering. Then I moved back to UC-Irvine as a Professor of Physics, where once again I had some interaction with Walter, but mainly on a personal level.

When Walter was awarded the Nobel Prize, it was not a surprise to me. The only surprise was that it was in Chemistry. This was quite logical, but my lingering impression of Chemistry dates back to my undergraduate days at University of Toronto. Perhaps he also endured these labs?

I learned technique and perspective in mathematical physics from Walter. He was my teacher and personal friend for more than 50 years. We have both reached an advanced age and continue to be functional. I congratulate him on his 80th birthday and I am most grateful to still be around to help celebrate this joyous occasion.

**About the Author**: Norman Rostoker is a Professor Emeritus and Research Professor in the Department of Physics and Astronomy at University of California-Irvine. He has been a Professor for 40 years, during which time he supervised to completion 40 Ph.D. students. His research since 1956 involves Fission Reactors, Controlled Thermonuclear Fusion, and Plasma Physics. He currently works on a source of energy, which involves the nuclear reaction of Hydrogen and Boron-11. The reaction products are three Helium nuclei. The radioactivity is negligible and Helium has no chemistry. The work is supported by the Tri Alpha Energy Corporation. From 1950–1953 he worked as a post doc for Walter Kohn and was a co-author of the papers on the K.K.R. method to calculate energy bands on periodic lattices. Contact: Physics Department, University of California, Irvine, CA 92717, U.S.A.; nrostoke@uci.edu

# It Started with Image Charges

Joseph Rudnick

University of California, Los Angeles, U.S.A.

I arrived at the University of California, San Diego as a graduate student in 1965. Walter Kohn's work with Pierre Hohenberg on the density functional method had been published, and I think that his first paper with Lu Sham on ground state calculations was also in the literature. I read their results on excitations as a typescript. I had my first encounter with Walter after I had completed one quarter of course work. I showed him my grades, which he pronounced satisfactory, and I began to work with him. I had approached him on the recommendation of my father, also a physicist, who described him to me as one of the best solid state theorists active in the world. His estimation of Walter was reinforced by Julian Schwinger a little later. I met him at a gathering and introduced myself as a student of Walter's. Julian's response was, "Ah, my most illustrious student." I knew that my future in physics lay in theory, as my only "D" ever was awarded for performance in the senior level experimental lab at Berkeley. In fact, theory was fine with me; I had no desire to go into competition with my father, a low temperature experimentalist of some repute.

In my second year of graduate school I was able to move into an office with Walter's stable of graduate students, which at the time consisted of Phillip Tong, Chian Young and Bill Butler. Eventually, Phillip and Chian moved on and Amit Bagchi and Bernie Nickel joined the group. This was a uniformly congenial collection of people, and some of my best times were spent their company. At one point, there was a call for a general strike on campus, in response to some issue or other. We agreed that we would have to put up a sign identifying our actions as a strike, as there was no other way of telling whether we were working or not. I regret that I have not kept up with my office-mates as I should have. Perhaps this volume will allow me to see what they are up to, from the abbreviated CV's if from nothing else.

After a couple of meetings, Walter set me my first project: to find a quantum-mechanical basis for the image charge. I think that he was interested in seeing how the image charge fit in with the density functional method. I was to reveal the connection between the image charge and solutions to

Schrödinger's equation for interacting electrons. Needless to say, I set about solving the problem in exactly the wrong way. It is embarrassing to recall the number of months I spent trying to pry the secret of the image charge out of the mathematical structure of non-relativistic quantum mechanics. It took the casual comment by Don Fredkin (at one of Walter's command-performance luncheons) that the image charge was the inevitable consequence of perfect screening to set me on the right track.

The question of the image charge and general features of metallic response to external perturbations at and near the surface of a jellium-model metal became the general topic of my dissertation. In the meantime, Walter and Norton Lang pursued their celebrated investigation into the surface energy of simple metals, carrying out the program that he was no doubt envisioning when he asked me to find out what I could about the image charge.

My interests have, over the years, strayed from electron physics. About three years after leaving graduate school I was introduced to the renormalization group approach to phase transitions, and ever since my obsessions have been largely confined to "$\hbar = 0$" physics. Nevertheless, Walter's spirit pervades my research, if his high standards are not always manifest in my output. I recall his comment when I mentioned that I was struggling through Abrikosov, Gorkov and Dzyaloshinskii. He expressed admiration for the abstruse and technically challenging calculations they performed. He then said, "I don't do that kind of work. I only do simple physics." It took me a while to appreciate this statement as the closest I'd ever hear him come to boasting. In the end, that is what I find most striking about Walter's work; the crystalline simplicity of it. The essential physics reveals itself, unembellished and pure. The Hohenberg and Kohn argument for the density functional method is astonishing in its austerity and power. Because of Walter, my highest goal is elegance and accessibility.

As a graduate student I never felt comfortable enough in the company of Walter to fully enjoy his affability. There was always a bit too much of hero worship – and accompanying fear of disapproval – to allow for that. However, I can boast of having played a small, but not insignificant, role in his life. When I was in graduate school I dated and eventually married Alice Cook, another UCSD student. Alice roomed with Naomi Schiff, first in a dorm suite and then in an off-campus apartment, at both of which places I spent a good deal of time. A few years later Naomi's mother, Mara, made the acquaintance of Walter. Upon learning that he was on the faculty at the University of California, San Diego, Mara asked her daughter if she had ever heard of a physics professor named Walter Kohn. Naomi had, of course, and gave her mother what was apparently a good recommendation, based on what fell from my lips during my many visits. The rest is history. Mara has told me that my high opinion of Walter played a role in her decision to allow a relationship with him to progress. I am happy to believe that this is so.

Joseph Rudnick

**About the Author**: Joseph Rudnick is on the faculty in the Department of Physics and Astronomy at UCLA. His current interests lie in condensed matter physics, particularly physics relevant to biological systems at the molecular level. Contact: Department of Physics and Astronomy, UCLA, P.O. Box 951547, Los Angeles, CA 90095-1547, U.S.A.; jrudnick@physics.ucla.edu

# A Math Teacher's Little Poem

George Sanger

Coronado, U.S.A.

The need for every Physics Major's path
is competence and skill in basic Math.
I was his teacher when he was 16.

The Nobel Prize he was not yet to gain,
but he applied himself, was bright and keen
in subject matter others found so weighty
I'm glad to notice now that he is 80
that all my efforts have not been in vain.

**About the Author**: George Sanger was born in Berlin, attended the Kaiser Friedrich Schule, and the Aberdeen University, Scotland MD, University of Toronto, Medical School. U.S. Navy, Reserve, Medical Corps. Private Practice. OBS/GYN Maine, California. Consultant U.S. Naval Hospital, San Diego, CA. Presently retired. Contact: 515 First Street, Coronado, CA 921108, U.S.A.; gsanger1@san.rr.com

# Stories About Walter

Andreas Savin

CNRS et Université Paris VI, Paris, France

Walter has many stories to tell, and many stories can be told about him. I feel reluctant to publish stories that are of such a personal nature. Knowing, however, the enjoyment I have in hearing stories about him, I tried to find a compromise and decided to write down just a few.

## Stories and Science

When Walter tells a story, he sometimes seems to leave the subject, only to return, later, with an unexpected twist. In science, he sometimes follows a similar pattern. One day he seemed to quite ignore something I told him, only to return the next day to the same subject, with a surprising and refreshing view.

## How a Proof Should Be

I found a proof and showed it to Walter. He said "one has to think more about it", so I checked the mathematics. Finding no error, I asked Walter to explain what he meant by his comment. He replied that he was not criticizing the mathematics but thought that one must strive to find the beauty behind the proof.

## Walter Judging Himself

Walter's publications are known for quality, not quantity. Indeed, some of Walter's work remains unpublished, because he finds that more work is needed to polish it. People around him knew this, but I, as a newcomer, was surprised to find that one of the subjects I was working on was in fact close to something Walter did 20 years ago and never published.

# A Technical Question

After much work, both numerical and analytical, a colleague and I found out that the Hohenberg-Kohn theorem is right, but that the one-to-one correspondence between the potential and the density is not "numerically true": a small change in the density can produce significant changes in the potential. I gave Walter an example for it: perturbing the potential by a rapidly oscillating potential. He answered: "This is known for a long time: a particle does not see the rapid oscillations in the potential."

# Walter About Motivation in Science

At a conference, I asked Walter if he understood the point of an investigation that had just been presented. He seemed to share my lack of interest, saying that the speaker can do with his life what he wants.

# Walter's Disagreement

Walter is known for being a careful listener. I rarely have seen him attending a talk without asking a pertinent question afterwards. There was, however, one exception. The speaker was only presenting slides full of routine numerical results. Once Walter realized this, he just closed his eyes.

# Is Walter Still Productive?

I met someone who knew Walter only by name. He told me that Walter must be unproductive by now. This was so surprising to me that I at first did not understand what he meant. It was easy to convince my interlocutor that he was wrong, by just quoting a few of Walter's recent papers. I hope that Walter will continue presenting us his ideas and surprising us for many more years.

**About the Author**: Andreas Savin is researcher at the Laboratoire de Chimie Théorique du Centre National de la Recherche Scientifique et de l'Université Pierre et Marie Curie in Paris. His current research includes fundamental aspects of density functional theory and the quantum mechanical analysis of the chemical bond. Contact: Laboratoire de Chimie Théorique, Université Pierre et Marie Curie, UMR 7616 du CNRS, Tour 22-23, 4 place Jussieu, F-75252 Paris Cedex 05, France; andreas.savin@lct.jussieu.fr; www.lct.jussieu.fr

# A Special Reunion

Douglas Scalapino

University of California, Santa Barbara, U.S.A.

One day Walter and I boarded a flight from Santa Barbara to San Francisco and sat together. Walter was headed to Washington, DC for a high school class reunion. I marveled at this and asked what year. Then Walter told me that this was the class of '40 and indeed a very special class. At that time, with pressure coming from the $3^{rd}$ Reich, Vienna had separated those students with Jewish names from the other students and they were sent to a different school. Walter had a class picture that showed the faces of a dozen or more eager boys and girls, posed for their senior photograph. This was the class of '40. Walter began to name them, recounting that this boy had gone on to become a well-known professor of mathematics, this girl had become a physician, this person an outstanding pianist and so on, truly a remarkable group of individuals. Of course, at the end of their senior year, the horrible chaos of that time would engulf them, sending them in all directions. Now, however, through chance and some searching they had found each other. I believe that four members of this special class were to attend this reunion, and Walter clearly was looking forward to it.

I was struck by the positive memories Walter had of what must have been such a difficult and frightening time. I asked about the teachers. Walter said that they were very special individuals. His Greek teacher also taught mathematics and physics, a devoted man who loved Greek and had his Ph.D. in mathematics. At that time, it was difficult for Jewish professors to find positions at universities so he was teaching at the high school. When the separation of the class had occurred, he had been assigned to be their teacher. He clearly taught them well and Walter went on to tell me many more tales about his teacher and this very special class.

I came home from my trip and wrote my children all about it. However, I can't find a copy of my letter and this was some years ago. Thus, I know that what I have written here is not precise, but the essence of that conversation and the indomitable spirit and joy of life of the man who recounted it to me has remained a very special memory I have of Walter.

**About the Author**: Douglas Scalapino is a professor of physics at UCSB. His research is in condensed matter theory. He first met Walter Kohn in 1966 when he spent six months at UCSD. With J. Hartle, R. Sawyer, and R. Sugar, he convinced Walter to come to UCSB in 1979 to be the first director of the ITP. Contact: Physics Department, University of California, Santa Barbara, CA 93106-9530, U.S.A.; djs@vulcan.physics.ucsb.edu

# Walking and Talking in Berlin

Matthias Scheffler

Fritz-Haber-Institut der Max-Planck-Gesellschaft, Berlin, Germany

We (i.e. the FHI[1] Theory Department) decided to present a joint *kohntribution*. I will start with some of my memories, leaving out, however, many aspects to keep things short and to avoid repetition with the preface of this book and the six brief personal reports from group members that will then follow below. A summary of our *tribute to Walter Kohn* may read: For us Walter is an icon, personifying the privilege to do science, how important it is to learn, how important and interesting it is to understand societies, cultures, and human interactions, and how much fun life can be.

It was only seven years ago, at the APS March meeting in St. Louis (1996), that I met Walter for the first time. (Since then he has been in Berlin quite often, and I was several times in Santa Barbara.) There he was: The man, I had told my students about, had influenced modern condensed-matter theory so strongly and at so many places that he deserves the Nobel Prize in physics for his lifetime achievement.[2] However, "the lifetime achievement" does not qualify for "the prize". There must be a singular topic. No doubt, density-functional theory is such a topic, and, for example, it is at the heart of the "total-energy workshop" series that we started in Europe in 1984 and which was later complemented by the $\Psi_k$-family[3], with Volker Heine the initiator, driving force, and father-figure. Meeting and talking with Walter in St. Louis was so easy and my initial nervous respect and admiration was gone after a second – just the nervousness was gone; the respect and admiration even grew.

In summer 1996 Walter visited Berlin. Nearly two years earlier the American Allies had left and handed back the "Harnack House", that they had

---

[1] FHI = Fritz-Haber-Institut der Max-Planck-Gesellschaft, also called "the Fritz".

[2] In 1998 Walter won the Nobel Prize in chemistry, not in physics, which is also an interesting aspect, but I will not dwell on it here.

[3] $\Psi_k$ is the tag of a European network fostering electronic-structure calculations (developments of new methodology as well as applications).

confiscated in 1945, to the Max-Planck-Gesellschaft (MPG).[4] Because the Harnack House is just across the street from the institute, I was temporarily appointed responsible. Thus, we put Walter into the "Gästehaus" of the Harnack House which we ran like a hotel, though with hardly any service.

German science history is clearly perceptible in Berlin-Dahlem, in particular in the FHI (as this is one of the two oldest KWG/MPG institutes) and in the Harnack House. Both places are now great, modern, active scientific institutions, but I also look at them as "warning memorials". Harnack[4] had always emphasized the international character of sciences. He argued that the KWG needs, what he called an "institute for foreign guests", and in 1929 such an institute was officially opened. Nobel Prize winners and their students met here in social exchange and for academic discussions, holding lectures and colloquia. All the big names went in and out: Albert Einstein, Peter Deybe, Werner Heisenberg, Fritz Haber, Adolf Butenandt, Otto Hahn, Lise Meitner, Otto Meyerhof, Max Planck, Max von Laue, Otto Warburg, etc.

In its first 4-5 years of operation the Harnack House was a great place. Then, however, the lecture program was increasingly affected by the Nazification. The "Kaiser Wilhelm Institute for Anthropology, Human Genetics and Eugenics", that was situated just 50 meters away, changed its direction of research rather early, building a "scientific background" for the terrible Nazi philosophy. They held lectures and courses at the Harnack House, and unfortunately the Dahlem scientists were not sufficiently sensitive to recognize the danger, or to stand up against it. Many scientists left Germany in the thirties.

And one other subject may be mentioned when talking about the Harnack House. In 1942 an important meeting of the researchers working in the German "atom project" and Nazi military took place here. Heisenberg was asked how long it would take to build an atom bomb and answered honestly: 3 or 4 years. The Nazis felt that this was too long and that the war should be over before then. Thus the development of a bomb was postponed.

It is still not understandable, and it is unimaginable, how things could change so fast in the 1930's, starting from an initially so very good concept and a learned atmosphere. Walter and I discussed it, obviously he with his personal experience, I only from reading. We also discussed about what was known about the Farm Hall transcripts and the skepticism Bohr and others had with respect to Heisenberg.

Dahlem is a "science colony". But, of course, Berlin is much more, and Walter and I have been to several places, for example at the Ludwigkirchplatz which is the area where his wife Mara had grown up. In 1996 Walter visited without Mara, but he wanted to see the area where she had lived until spring 1939. Seeing kids playing was so nice and peaceful, but we also knew that some of these benches had been painted yellow in Nazi Germany and where marked "nur für Juden". Neither Walter nor I had

---

[4] Adolf von Harnack was an initiator and the first president of the Kaiser-Wilhelm-Gesellschaft (KWG) which later became the MPG.

mentioned this, we just were happy seeing the present scene. I had similar feelings when Walter and I walked from the "Staatsoper Unter den Linden" along the museum island to Oranienburger Straße, visiting the synagogue, which is now a Museum. We also walked through the Hackesche Höfe and back to have a Jewish/Arab dinner at Cafe Oren (next to the synagogue). Here we talked about a conference I had attended in China four years ago, and we also talked about scattering of noble-gas atoms at surfaces and the incorporation of van der Waals interactions in new $xc$ functionals. There are so many more things I do remember from 1996 and the various later visits (for example, a concert in the Philharmonie, another one in the Centrum Judaicum, dinners at our house together with my wife Barbara). Many nice moments, some experienced even without words, though Walter always has a story to tell. His life is loaded with so much experience. Also to mind come my feelings when we went last year (2002) through the Tiergarten (see the photo of the Siegessäule and the story by Cathy below). I was walking with Mara and seeing many different ethnic groups (Turkish, Iberian, Korean, German, etc.) having lunch or coffee, and seeing the kids and parents playing was again nice. It was so peaceful in the Tiergarten. Why can't we export this spirit, as well as good sense, why can't we cut down ignorance, egoism, and arrogance? This is not the place to comment on the world as it seems to develop right now, but I know, Walter's and my assessment would be similar. Also in his involvements in political and social issues Walter's competence, intellect, and talent to address a subject matter in clear words, brief, and still accurate is striking.

## ...Exact Solutions (Axel Groß, now at TU Munich)

In 1997, when I was working as a post-doc with Matthias, I had the pleasure of meeting Walter Kohn when he visited the FHI. In fact, it was not the first time I had met Walter. Ten years earlier, I was spending one year as an exchange student at the University of California at Santa Barbara and took the condensed matter class that Walter Kohn was teaching. Instead of a final exam, we had to write an end-term paper. I decided to write about the glass transition, and visited Walter in his office to discuss my assignment. Obviously, I was frustrated with the incomplete state of the theory of the glass transition at that time because at a certain point of the discussion Walter said to me: "You are a young physicist. Obviously you still believe that physical problems can be solved exactly." Indeed, only then I fully realized how important it is to find the appropriate approximations in the description of physical systems, a fact Walter had often emphasized in his class. Apparently I took Walter's remark seriously when writing the topical paper because he graded it with A+ which I am still very proud of.

Being now a professor myself, I always tell this little anecdote of Walter Kohn in my quantum mechanics class when I start teaching approximation methods. I always feel that this helps the students to appreciate the importance of perturbation and variational methods in physics.

**The Social Person** (M. Veronica Ganduglia-Pirovano, now at Humboldt University, Berlin)

In 1997 Walter visited the FHI for a week, and Osvaldo Rodriguez (from La Plata, Argentina) was there at the same time. Matthias came to me and said "Could you please take care of Walter K., I can not", and he explained that there where important personal reasons. As it is usual when we have guests, a dinner was already planned. This time at a Thai Restaurant in Prenzlauer Berg. During the dinner I told myself: This guy is truly special! He would tell stories, all kind of stories! The topics would vary from politics to geography, different places, customs, culture, people but NO science! That night was a social event and not everybody at the table was a scientist. Moreover, after the dinner and already on the street (full of bars) people were little hesitating about what to do next and mentioned the interesting area and its possibilities .... The next I remember is that Walter and most of us were at a bar close by where the conversation continued as lively as before ... and still no science.

Entertainment for the next day was not planned since under normal circumstances Matthias would have taken care of such a special guest. Actually most of us had planned to attend a birthday party of a Ph.D. student. We mentioned this to Walter and invited him to come along. And there he was, the next day, in the middle of the mostly young crowd at the birthday party. Osvaldo said: "Walter fitted in the group extremely nicely, sharing the party as another friend". Indeed, he would talk lively as time would not be a variable for him, always smiling and extremely interested in other people's points of view. – To take care of such a guest was a reward for us!

**Introducing Walter** (Catherine Stampfl, now at University of Sydney, Australia)

Due to the notoriously busy travel schedule of Matthias, I was to have the honor (and responsibility) of introducing Walter Kohn at his seminar at the FHI in summer 2002. Having only exchanged a few words with Walter at the Mensa (student cafeteria) when some of the Theory group escorted him for lunch when he visited some years earlier, I was sure he would not remember me.

Walter had arrived, and later I was called into Matthias's office to join them for lunch. My first surprise came when walking to the restaurant, Walter remarked that he remembered me; well, he remembered there was someone who looked like me and who was from Australia. On observing Walter Kohn, and later talking with him, he impressed me first as being a very happy person. One with an ample and keen sense of humor, with a smile never far from his

Aerial view of the Tiergarten and the Siegessäule monument, which nowadays is more a symbol of peace and love, and for example the final point of the "love parade" that attracts about a million "Techno fans" every year (© ullstein bild, Lothar Willmann)

lips. He also seemed to have a knowing twinkle in his eye, a genuine interest and excitement in life and all its wonders.

On Saturday, we took Walter and Mara on a boat trip (see the photo in Dan Hone's contribution), and then a walk through the Tiergarten (Berlin's "central park") to the Siegessäule (see picture) which has very many internal spiraling steps to climb up, so that at the top, one has a lovely view of Berlin. This was actually a rather long walk. In any case Matthias had planned to offer a Velo taxi to Walter and Mara, but both were determined to continue walking; and moreover, on getting there, Walter climbed all the way to the top (see picture), whereas some younger colleagues thought they didn't have the energy.

The time came for Walter's seminar. The room was over full. I managed the introduction, not without some nerves. Needless to say the seminar was very entertaining and interesting. For the evening I had arranged for a "Nachsitzung"[5] in a local beer garden and we had nice weather. There were about 20 people ranging from senior scientists to post-docs to students, all arranged at one big long table with Walter and Mara in the center. Indeed it can be somewhat perturbing to be faced with 20 eager faces peering at you; some of whom Walter would know, but many who he would not. I was impressed by Walter's action; he requested that everyone in turn should introduce them-

---

[5] "nachsitzen" is used as a pun in German. It refers originally to a punishment of children who have to stay in after school due to some wrongdoing. Here, the term is used ironically, referring to *sitting* down for a social dinner *after* the lecture.

*Left*: Walter at the top of the Siegessäule lookout; *Right*: Walter climbing down the stairs inside the Siegessäule (summer 2002, Berlin; photos: M. Scheffler)

selves, saying their name and mentioning a little about what they were studying. After this, conversation flowed freely and a very nice time was had by all.

## Walter and the Music (Klaus Hermann)

On his recent visit to the Theory Department I found Walter in his office trying to find cultural events worth visiting on the weekend. He was also interested in music and I suggested a concert in the world famous Berlin Philharmonie. On that weekend the Berlin Philharmonic Orchestra was performing a Shostakovich symphony and Berg pieces and Walter was a bit reluctant that his wife might not like the music. Besides, I found out that the concert had been sold out a long time ago. So I suggested a concert of the Berlin Philharmonic Chorus performing Haydn's Four Seasons in the Philharmonie. Walter was excited immediately since this music piece had been very popular in his family when he was still a young boy. After a short consultation with his wife he asked me to get tickets. During the concert I could watch Walter and his wife from the stage, where I was singing in the tenor section, and there they were sitting near the center of the hall paying close attention to the music. A few days after the concert I asked Walter whether he liked the performance. "Oh yes, my wife and I had a wonderful time in the Philharmonie. We enjoyed the music very much." Then he put on his wise smile and said "You know, we even had a lovely conversation with the couple sitting next to us. It turned out they were real Berliners."[6]

## A Remarkable Experience (Martin Fuchs)

It is a nice tradition in the department that visitors are brought in contact also with students. Thus, I had met Walter Kohn the first time in 1997 as an undergraduate and again in 2002 as a graduate. I certainly did remember how I had felt after a previous discussion. There I had made a remark on how reliably certain density functionals would describe the compressibility of solids, and Walter Kohn vividly objected that my thinking was seriously in conflict with other physical arguments he had written up already decades ago, and which I had been ignorant about. The argument was left unsettled and, needless to say, this was a quite sobering experience: I had not exactly planned to impress him as a "naive". Clearly, my motto for the second encounter in May 2002 was "better be better prepared". This time, he was explaining his views on van der Waals interactions and I had some related results. I was impressed, though again growing a bit nervous, by how he, not doing numerical computations himself, very crisply and in much detail pinpointed many of the numerical hurdles and pitfalls that I was struggling with. Our meeting went on for quite a long time and at the end I felt somewhat exhausted. Less so Walter Kohn. He was already looking forward to meeting his wife and scheduled to attend a reception somewhere downtown Berlin. I suggested to him to relax and take a taxi, but he was very curious to explore, on an extraordinarily hot day in late May, the Berlin subway system and eagerly asked for directions. Indeed his visits are "A remarkable experience!" he expressed the next day.

## Long-time correlations or How to Save a Balcony (Arno Schindlmayr)

When the FHI Theory Department had finally reached the limits to growth within its traditional departmental building, new space for expansion was found in the Richard-Willstätter-Haus, a recently vacated former directors' villa on the institute campus. After overseeing the necessary construction work, I moved into the new offices together with my Ph.D. students and co-workers in the spring of 2002. A few months later, Walter Kohn was among the first visitors we welcomed here. Due to the fine weather, we settled on the balcony, a towering spacious platform overlooking the garden that had been erected at the instigation of the last resident director in the villa, and which had immediately become popular with us, the new inhabitants, as a place for recreation, while nonresidents perceived the looming modern construction almost unanimously as an architectural insult to the eye and the historic integrity of the villa. Despite an appeal, the balcony was hence already scheduled

---

[6] At that time, the Theory Department included only one genuine Berliner, Matthias Scheffler, and Germans formed a minority.

to be returned to its modest and significantly smaller historic appearance. As it happened, however, Walter also visibly enjoyed the lofty place, and in the absence of other commitments we spent a long sunny afternoon outdoors discussing fundamental issues of density-functional theory and some surprising manifestations of correlation. The attentive crowd around Walter was eventually noticed by passing decision makers at the helm of the FHI, and shortly afterwards, without any further intervention, we were informed about a revised decree that keeping the balcony was in the best interest of the institute after all. It thus survives due to wondrous long-time mechanisms actuated by this visit, and I would like to say thank you, Walter!

# My Personal Walter Kohn Story

Sheldon Schultz

University of California, San Diego, U.S.A.

It is my pleasure to have been asked to contribute some reflections about my colleague and friend, Walter Kohn. As I reminisce, flashbacks come to mind and although they may individually not be the most important, perhaps collectively they compose a personal recollection portrait.

George Feher who was to become my mentor, colleague and friend, came to Columbia University as a Visiting Professor while he was at Bell labs. He had been asked to help nucleate the newly-developed Physics Department at what was going to be a major new campus of the University of California multi-campus system then known as UCLJ (yes – University of California La Jolla), how that name changed to UCSD is another story. I had recently completed my PhD defense under the sponsorship of Professor Polykarp Kush on atomic beams, but had announced my intent to go into solid state physics as it was then called. Modern solid state physics had been a forbidden topic at Columbia because another Nobel Laureate (I.I. Rabi) had ruled that it was not important compared to the blossoming field of high energy physics (T.D. Lee had also just received his Nobel award). As I later learned, George asked Kush to recommend an energetic, competent (I hope) young postdoc who might be recruited to help set up a new laboratory at UCLJ where George was planning to go. Within a month, I was flying to La Jolla to interview and where I met Keith Bruckner (the first Chairman of Physics) and Walter Kohn. Even in that brief early experience I sensed that Walter was someone I could relate to, and indeed, that intuitive feeling was amply justified by subsequent experiences.

Not knowing any solid state physics (literally!) it took time for me to fully appreciate the work that George, Walter, and Harry Suhl had already accomplished (much less, could I appreciate the alchemist Bernd Matthias). But I soon came to realize the impressive collection of theoretical post docs, visitors, and students that Walter and Harry brought in those earliest years. This brings me to the only professional memory that I will mention: Walter's weekly informal Theoretical Solid State Lunch. Although this was open to all faculty, staff, postdocs, and students who wanted to attend, I was the only experimentalist who regularly did so. I originally felt I needed to attend as part of my crash course in what solid state physics was all about. Eventually I became the token solid state experimentalist, and I must confess that I thoroughly enjoyed playing that role. The general discussion of a wide range of topics during lunch, and the informal reports that we all made after, are something I have missed ever since Walter left for Santa Barbara. I still miss it.

Walter was an excellent example of my own image of what a Faculty Professor should be. A dedicated scholar, teacher, an active contributor to the workings of the campus, and in general a person of social conscience. I will not refer here to details of his soon becoming Chairman of the growing Physics Department, nor his later Chairmanship of the UCSD Academic Senate. But I will flash forward to the mid 1960's when we had all already moved (out of the Scripps Institute of Oceanography at the beach) up to the new main Campus being built on the mesa. It wasn't long before many emerging social issues became dominant topics on the campus – (this was pre-Vietnam) – critical issues such as visitation rights, and why we had all-male and separate all-female dormitories, and of course, the poor quality of dormitory cafeteria food. Why do I mention such trivia? – because when a food fight broke out spontaneously one night and rapidly escalated to a *cause celebre* it became important to deal with this issue, particularly when the food fight issue was being led by a young man who became the spokesman of the dorm students.

Professor Kohn was serving on yet another faculty committee and in an effort to help the provost/administrator of the first college (soon to be called Revelle College), they convened a student meeting to hear all the food grievances. I have to mention that Walter had lost both his parents in the Holocaust and as a young high school student escaped to England, and then to Canada and finally to the U.S. This young student leader was himself Jewish, and should have had more sense because when he spoke about the "suffering" his fellow dorm students had endured, he likened it to the medical experiments that Nazi doctors had performed on Jewish prisoners during those horrible years. Why do I bring up such a painful topic? Because Professor Kohn (in modern language) somehow kept his cool, made some appropriate observations, and handled himself in a way that made a deep impression on me.

Lest the reader think Walter was not a man to take action, reader beware – because when I listed as my other criteria, that I expected a faculty professor (to be a "person of social conscience") it soon became clear to me that Walter was to take up many causes that I also believed in, and he pursued them with passion and dignity.

And so I conclude with a toast to Walter on his 80$^{\text{th}}$ birthday, by saying what my grandfather would have appreciated if he were here to listen – "Zadeh – he is great scientist and a Mensch."

P.S.: Everyone who knows me, very quickly calls me Shelly (totally unknown persons send me emails addressed to "Shelly"). So I must confess how amusing I find it that from Day 1 Walter has called me Sheldon. I have many times said to mutual friends that Walter is the only person to call me Sheldon since my mother called me that in my youth but only when I was in trouble: "Shel-don!" Please keep doing so Walter – to me it's part of your special signature.

**About the Author**: Sheldon Schultz has been a member of the Physics Department of UCSD for over 42 years. His current interests are the new Negative Index of Refraction meta-materials, and plasmon resonant metal nanoparticles for bio-medical diagnostics. Prior areas include magnetic nanoparticles for information storage, microwave spectroscopy of spin glasses, High-$T_c$, and conduction electron spin resonance in metals. Contact: Department of Physics, University of California, San Diego, 9500 Gilman Dr., La Jolla, CA 92093, U.S.A.; sschultz@ucsd.edu

# Walter Kohn and Vienna

Karlheinz Schwarz

Vienna University of Technology, Vienna, Austria

## The Meaning of KS in Science

As I look back on my own career and my relation to Walter I find that science often takes unexpected turns. After my PhD in chemistry at the University of Vienna I had the opportunity to take a postdoctoral position in J.C. Slater's group at the University of Florida in Gainesville in the years 1969–71. In his group many discussions were held on how to treat exchange in extended systems. For me the starting point was predefined with Slater's exchange but the pre-factor, whether it should be 1 or 2/3 or something in between, was heavily discussed for example at the Sanibel meetings. This was the time when the $X\alpha$ method was introduced and where I became "mister alpha". What has that to do with Walter?

It was many years later that I learned about density functional theory, with the famous Hohenberg–Kohn and Kohn–Sham papers from 1964 and 1965. Therefore in the early seventies these papers were around and available but I must admit that I was not aware of them and did not realize the importance of this work. Nowadays I may find an excuse arguing that I started as a young researcher from the other camp. Several years ago I discussed this matter with Walter, who was not upset that his important work was not fully recognized in the early days but he simply told me that new ideas take time to be accepted.

When Walter received the Nobel Prize in chemistry rather than in physics and prepared his Nobel lecture for Stockholm he asked me for an important application in chemistry. I was more than pleased to learn that eventually he had chosen a picture from our work as his first figure, namely the ab-initio molecular dynamics simulation of the catalytic reaction of methanol in a zeolites, which was based on DFT and made it to the cover page of "Angewandte Chemie" (Applied Chemistry). For many researchers in our community the solution of the Kohn–Sham equations is our daily basis and KS stands for it. This is often true for me as well but in my case there is an additional meaning of KS, since these letters are my initials, which makes me proud. I had many discussions with Walter on DFT and related topics, which were extremely fruitful for me, but they were – and still are – always in English. This continued even after the time when I realized that he was born in Vienna. Being a Viennese myself, this sometimes feels strange but considering the dark period in the Austrian history and the fate he and his family have experienced I fully understood and accepted his reaction. Under these unfavorable circumstances it means even more to me that he accepted me as his friend and continues his friendship. Over the years we slowly began to talk about areas outside science and we started to speak in German. In this case it was more than just another language. It gave me the opportunity to see Walter from a new side, namely as an outstanding personality.

## Walter and the "Akademische Gymnasium"

At more than one occasion Walter told me his reservations and his experience with respect to Vienna that nicely characterize his personality, and so I'd like to tell it here. The main steps in Walter's life are well known by now and thus I can be brief mentioning only some crucial points. When the decision had to be made in the Kohn family to which school Walter should go, it was clear for his Jewish parents that it had to be "Akademisches Gymnasium", which was onsidered "the best". In 1938, however, when Walter was fifteen years old, he had to leave this school as a consequence of the Anschluss to Nazi-Germany. What followed can hardly be described and will remain in our memories as the dark period in which our country has lost so many outstanding personalities who contributed so much to the cultural world. This is not the place to write about it.

About sixty years later Walter was back to Vienna. At this time the school, the "Akademisches Gymnasium" again, asked him, the Nobel laureate, to visit "his" school. It was quite obvious that the school wanted to show off with its famous son, with a Nobel laureate. Walter admitted that – at first – he was a bit upset but then he accepted and went to the school, where they showed him around. Among other things he attended a history class, in which the teacher had started a project on the Nazi-period looking at specific family tragedies. The pupils started conversations with Walter and asked him many questions about that time. It was these discussions that made him see Austria differently, in a way that was new to him. Now comes the unexpected reaction of Walter. He changed his mind and made a significant donation to the school in order to continue such efforts. It takes a great personality to donate so generously to the school that once expelled him. This requires the highest respect and shows Walter's outstanding personality that we all admire.

## Celebrations

In 1997, one year before Walter's Nobel prize, I organized a DFT conference that was well attended by the ever growing DFT community from physics and chemistry. Walter gave the main lecture and we had an excellent conference with lively discussions. My wife Christine helped me with the organization and she made it possible to arrange the conference dinner at the Natural History Museum. We first had a look at the exhibits such as the famous Venus of Willendorf, a statue about 25 thousand years old or at the mineral collection. Then we had our dinner on the first floor, where the tables were arranged under the cupola with a view of the beautiful ornamental marble decoration in the ground floor. It was a warm September day with a clear sky and the splendidly lit Ringstrassen buildings. Looking through the windows one could see the famous Museum of Fine Arts in shining light and the statue of Maria Theresia between the two museums. Across the Ringstrasse the renovated dome of the Hofburg with the green cupola, its golden decoration and the Habsburg crown on top of it could clearly be seen. The main buildings of the Hofburg and the Heldenplatz appeared in warm evening light. One of the participants called this an "Imperial Dinner". Late in the evening Walter thanked in particular my wife for this outstanding dinner. This was a day when we could show to the world the beauty of Vienna and we were proud of being Viennese.

Four years later my group organized – as sort of a surprise – a special conference, Applied DFT2001, to celebrate my 60th birthday. A number of my friends and well-known scientists from all over the world were invited and they all came on this occasion. Walter accepted to be honorary chairman and to give the opening lecture, the highlight of this event. The members of my group started to find sponsors for this conference and found among others the city of Vienna, who wanted Walter to give an official lecture, for which they were willing to finance his travel expenses. This was nicely planned, but the situation changed when a new government was formed in Austria, in which Haider's Freedom Party participated, a party that can be described as a populist right-wing party. Under these circumstances Walter did not want to have anything to do with the official Austria. Consequently he refused to give any television or radio interview let alone an official lecture, activities that were originally planned. However, friendship is something different. He had promised to come to my birthday and so he did and gave a splendid lecture on van der Waals interactions. Furthermore, he contributed to the discussions of almost all the other talks and thus made this a remarkable conference not only for me. Just imagine such a birthday present. It is fascinating to see and experience how Walter appreciates his Austrian friends on the one hand and on the other is very critical of certain political trends in this country. He will be pleased to read that in the recent election the Freedom Party lost two thirds of the votes.

During this four-day event the evenings were devoted to celebrations with lots of laughter. Once we were sitting at a table in a restaurant and Joachim Luitz, one of my co-workers, brought a little boat that was running around in a small bath-tub, but obviously without any engine. The lecturers – among them Ole Andersen, Peter Blöchl, Roberto Car, Helmut Eschrig, Dimitri

Vienna, January 16, 2001, celebrating the 60th birthday of Karlheinz Schwarz with his wife Christine and Walter Kohn (photo: J. Luitz)

Papaconstantopoulos, and others – had long arguments (with lots of laughter in between) on what makes the boat go round and produce a sound like an engine. Finally everybody had to write their explanation. Walter was asked to be the referee, read all the proposed theories and then he concluded in his report "read but not approved". It was a hilarious evening that all of us will remember.

**About the Author**: Karlheinz Schwarz is the head of the Computational Solid State Theory Group in the Institute of Materials Chemistry at the Vienna University of Technology. His research focuses on the interplay between electronic structure and properties of condensed-matter systems. The program package WIEN2k was developed in his group and is provided to a large scientific community. Novel material of increasing complexity can be studied by density functional theory calculations e.g. using this code. Contact: IMC, Getreidemarkt 9/165-TC, A-1060 Vienna, Austria; kschwarz@theochem.tuwien.ac.at; info.tuwien.ac.at/theochem/

# We Met at the Institute in Copenhagen

Benjamin Segall

Case Western Reserve University, Cleveland, U.S.A.

Shortly after arriving at the Institute for Theoretical Physics in Copenhagen (now known as the Niels Bohr Institute) in January 1952 at the start of a postdoctoral fellowship, I met Walter Kohn. He had already been there for about a half a year. The combination of my being abroad for the first time and being at that institution that was so renowned in the world of physics made me feel a bit overwhelmed. In his usual open, friendly and generous manner, Walter went out of his way to make me feel at home at the Institute, and ultimately to help make my stay there more profitable and enjoyable. In addition to talking to me about physics and other matters, he invited me to his home on several occasions . Considering that he and his wife were also away from home and that they had a young and growing family, the invitations were most generous. And, the warmth of the Kohn's made the visits memorable.

Walter's year at the Institute ended in the summer of 1952 when he moved to Pittsburgh to start a professorship at Carnegie Mellon University. I was sorry that he was leaving. The move back to the States was not going to be a simple matter for the Kohn's as they had their second daughter (Ingrid) just a few weeks before their departure and by that time had accumulated a fair amount of things. The trip back to the U.S. was to be by ship from Copenhagen. I went to the port both to say goodbye and to offer to help them load some of their bags and other items. The most special package that I helped with, the only one that I remember, was a basket containing the baby Ingrid. I was, in fact, proud that Walter entrusted me with such a precious package. Needless to say, it was handled with the utmost of care.

A few years after returning to the States I joined the staff at the General Electric Research Laboratory. There I started to work on condensed matter research and became interested in electronic structure calculations. Shortly after that I saw a paper by Walter and Norman Rostoker that appeared in the Physical Review on a new method for carrying out such calculations. Walter's name on the paper stimulated me to study the paper in detail. The approach they used was based on Schwinger's variational principle and was very neat and elegant. But more importantly, it gave promise to be more accurate and efficient than other methods then available. The formulation in the paper was restricted to crystals with just one atom per unit cell while I was interested in studying, among other materials, semiconductors, whose crystal structure involved two or more atoms per cell. Consequently, I undertook the task of extending the method to handle an arbitrary number of atoms per cell. When I wrote up a paper on the work, I requested that Walter read and comment on it. He generously agreed and made several helpful suggestions.

Soon after that, Frank Ham, who also was interested electronic structure calculations and who independently had started to work on the Kohn–Rostoker method while he was a postdoc at the University of Illinois, joined the G.E. Research Laboratory staff. With such mutual interests we joined forces. We proceeded to implement the method for what was then large scale computer calculations. It was the time that computers like the IBM 701s, power houses compared to their predecessors, came out. But more importantly, it was the time that significant data about the Fermi surfaces and the electronic structure around the Fermi energies of a number of metals were being revealed by a variety of techniques like the de Haas-van Alphen and cyclotron resonant measurements. Frank used the method to study the bands of the whole family of alkali metals. Meanwhile, I worked on the band structure of Al and Cu. The fact that our results, particularly those for Cu, were in fairly good agreement with the recently published Fermi surface data was very gratifying and exciting. I think it is fair to say, that that was the time that it was becoming evident that electronic structure calculations could yield interesting detailed results about real materials – even those that did not have "free electron-like" bands. At the time of our work, in order to perform the calculations we had to construct a reasonable crystal potential for a particular material, or, employ some other method to obtain logarithmic derivatives well outside the "core" – Frank Ham used the "quantum defect method" for the alkalis. (I am saddened to report that my good friend and old colleague Frank passed away in December). As we all know, the development of the density functional theory by Walter and co-workers provided a sound basis for electronic structure calculations. And that work provided a practical means for obtaining a well defined self-consistent effective potential. It hardly needs restating that that work, for which Walter received the Noble prize, was pivotal in making electronic structure calculations as powerful and useful as they now are. I like to think that the success of Frank's and my calculations, and, of course, similar successes of many other of the early "band calculators", may have had something to do with Walter's recognizing the need for providing a sound basis for the calculations.

In 1968 I joined the Physics Department at Case Western Reserve University which had just formed from the merger of Case Institute of Technology and Western Reserve University. Shortly after my arrival there, the chairman of the department requested that I arrange for a special event, sort of a coming out party for the newly merged department. It was to involve a week-long lectureship by an outstanding nationally known physicist. I immediately suggested Walter. The Chairman agreed; and shortly after that so did Walter. In the spring semester of 1969 Walter came and gave an excellent series of lectures on the range of subjects that he had recently been working on. All the lectures were well attended and very well received. During his weeklong visit, he met and interacted with almost all of the fairly large number physicists who were in the department at that time. Walter's lectureship was uniformly considered an overwhelming success by the faculty. That was a source of considerable satisfaction for me. And, of course, I personally greatly enjoyed his visit.

In subsequent years, my contact with Walter has, unfortunately, been more limited. These have mainly been at conferences, most often at the March APS meetings. But these short contacts have been especially pleasant and rewarding occassions, ones that I especially looked forward to. I am sure that most other physicists who had the pleasure of knowing Walter feel similarly.

**About the Author**: Benjamin Segall is Emeritus Professor of Physics at Case Western Reserve University. His area of research is condensed matter physics. His interests are in the calculation of the electronic structure and properties (mainly optical) of semiconductors, semiconducting alloys, superlattices, interfaces between semiconductors, and between semiconductors and metals. Contact: Physics Department, Case Western Reserve University, Cleveland, OH 44106-7076, U.S.A.; bxs2@po.cwru.edu

# Happy Birthday, Walter

Lu J. Sham

University of California, San Diego, U.S.A.

"Birthdays are good for you, Mr. Wilson. The more you have, the longer you live."

— Dennis the Menace, by Hank Ketcham, Jan. 20, 2003.

To the calls for contributions to this book to celebrate Walter Kohn's eightieth birthday, there will be an outpouring of fond good wishes, as there were for his sixtieth birthday and his seventy-fifth. Walter treats his work colleagues as friends and family. Walter considers Quin Luttinger, a long time collaborator, one of his best friends. When I was his postdoc in La Jolla (1963–66), he introduced the students of Quin Luttinger to me as cousins. T.V. Ramankrishnan was one who comes to mind. Indeed, from Walter's legion of collaborators, postdocs and students, I have made many friends, with Pierre Hohenberg, Norton Lang, Vinay Ambegaokar, Bill Butler (some of whom would no doubt have their own say in this book), even though we do not meet often.

Besides his renowned work on density functional theory, Walter Kohn has made many fundamental contributions to solid state physics. With Quin Luttinger, he provided a method for computing the properties of charge carriers in semiconductors which is used to this day. His theory of the shallow donor was recently celebrated since impurities are of course the lifeblood of semiconductor devices. He treated the extra electron in a semiconductor as the $(N+1)$th electron with interaction in the many-body system and showed that it could be treated as a single particle with some "renormalized" properties, so that the other $N$ electrons could be put out of mind. This argument becomes relevant again recently when it was challenged whether the semiconductor with its gazillions of electrons could be used for quantum computing or information purposes. From the same line of research came also the "Kohn Theorem" that the electron interaction does not enter into cyclotron motion (in the sense of a rotation of the system). This also comes to prominent consideration for recent experiments in micro- and nano-structures.

Walter and Quin also attacked the basic problem of the day, to prove the existence of the Fermi surface. He told me that they failed. Actually, on closer examination of their work, they showed that at very, very low temperatures, pure electron-electron interaction could induce superconductivity which meant that the Fermi surface could not be a ground state property.

Then there is what in polite company is known as the Kohn effect in lattice vibrations of metals. Kohn gave a characteristically simple but elegant argument that the electron response to the lattice vibration would change as the excited electron momentum increase exceeds the Fermi surface diameter. At first the experiment could be made to show the Kohn effect only if one examined the atom-atom interaction as a function of the distance to interpret the Kohn effect as a Fourier transform of the Friedel oscillations. As quasi-one and two dimensional metals became reality, the Kohn effect appeared directly in neutron measurements. In Walter's group in La Jolla in the sixties and seventies, the Kohn effect was always referred to as the Kohn anomaly accompanied by a big smile, for the thought of another Kohn anomaly was so impossible. (Really? Read on!)

With the founding chairman Keith Bruckner, Walter Kohn built the Physics Department at University of California at San Diego into a well-known one in a short time. I got a sense of this achievement from the victims of his raids of talent. In the eighties, our department recruited in succession three physicists from Bell Laboratories, Duncan Haldane, Cliff Surko, and Bob Dynes. Then I used to spend summers at Bell. One summer when I showed up, there was a clamor in the tearoom to give me a red escort badge, which meant that I had to be accompanied everywhere, including the bathroom. People were upset because Charles Tu had just left for San Diego also. I was then a dean but I had nothing to do with some of these recruitments. A friend who was at Bell since the fifties explained that people thought it was deja vu from the fifties, when after every summer visit by Walter Kohn, someone would move to La Jolla: Berndt Matthias, George Feher, Harry Suhl.

Walter has a larger effect on the San Diego campus. He was the driving force behind the start of the Judaic studies with an endowed chair. He was instrumental in setting up rules and regulations on the young campus. Right after he left in the late seventies for Santa Barbara as the founding director of the Institute for Theoretical Physics, I was chair of the Committee of Educational Policy at the San Diego campus. Every time (surprisingly frequent) the committee got embroiled in some hot campus educational affair, the supremely capable Lynn Harris would dig up a memo from the fifties which set a sensible precedent on what to do. The signature at the bottom of the memo would be the familiar flourish of Walter Kohn, chair, CEP.

Walter had a great interest in Jewish history. When we were in Krakow for a conference in June 1994, we went to visit a famous historic Jewish quarter north of the city. Walter gave my wife Georgina and me a detailed and delightful tour of the synagogue. We ended up having tea in a book store in the quarter (see photo on page 28). Upon Walter's recommendation, we bought a beautiful picture book, "Journeys to the Promised Land". The book was put together in Israel, published in London, and printed and bound in Hong Kong.

Walter is interested in politics. He is one of the few people I know who actually act on his belief. He is, for example, active in presenting arguments to persuade the regents of the University of California to discontinue management of the nation's weapons laboratories. About ten years ago, during my sabbatical leave in Santa Barbara, Georgina and I were invited by Walter to lunch at home with a few friends (which meant several tables in the garden). Mara was out of town and so Walter proudly announced that he would take care of lunch. Everything was going swimmingly until the main course. The subject of the regents approval of the next contract on the national laboratory management came up at our table. Incautiously I said that I thought the contact of lab personnel with the academic faculty was a good point of having the University managing the labs. Whereupon Walter audibly organized his thoughts and marshaled his facts and proceeded to treat us with a fact-filled and carefully analyzed account of why the University should have nothing to do with the weapons labs. By that time, all the other tables were waiting for the next course and Georgina was throwing looks at me which were not, as one comedian put it, the looks for which I married her. Walter calmly stood up and announced that he would serve a delicious (with a deprecating characteristic sound somewhere between a giggle and a laugh) dessert right after he finished this important lecture (dead earnest but the exact words are lost in the fog of history) to Lu.

Everyone who knows Walter must know the care with which he speaks. He is an interesting raconteur. But few people without personal experience would believe that he is a daredevil in sports. The first inkling came with a phone call from the hospital around March, 1965. Walter asked me if I could take his place for an invited talk (the first one on the density functional theory) at the March Meeting of the American Physical Society. He had broken his shoulder in several places in a bad fall while skiing. Since he was known as an excellent skier, I pressed for details. Well, on the wide slope in June mountain, he saw a man going straight down without traversing, flying over the moguls gracefully. That reminded him of his youth in Austria where he used to build ramps by hand for ski jumps. So Walter decided to follow suit. Oh yeah, he broke his leg the time in Austria too.

One summer in La Jolla after a storm, the wind was stiff and the surf was high. The surfers rushed to the beach. And Walter invited me to try out his new sailfish. (To this day, I do not know if the weather and the invitation were connected or not.) Both families came to Kellogg's beach. Marilyn (his eldest daughter) joined us sailing. We swam the sailfish over the surf with a little difficulty. The stiff wind did propel the sail of the supersized surf board rather nicely until a squall wiped us out (capsized the boat). After vain attempt at righting the sail, we started pushing the sailfish towards shore. The sailfish was ripped off our hands by the heavy surf and we had to swim past the rough stuff to reach land. We all sprawled on the sand while a lifeguard stood over us listing all the dangerous acts we committed. As I dozed off, he was saying that, for starters, the sailfish was built for two ....

There was the triumph of the soccer match at the first picnic of the physics department. The graduate students challenged the faculty to games. So a soccer and a baseball match were negotiated. The faculty was small and press-ganged the postdocs into service. The faculty members were very good soccer players. I was in the backfield with nothing much to do except admiring these old men (old is relative as I now find out), especially Walter Kohn and George Feher, running circles around the students. They seemed to have a simple strategy. They simply kick the soccer ball above the ground but with an arm's reach of the students. The graduate students grew up mostly playing American football in which no players except a designated one are allowed to kick the ball. So they instinctively reached out to bat the ball down. The faculty side won with the advantage of a rash of penalty goals. Then they were too tired to play baseball, which was just as well, from the point of view of preserving the soccer victory.

With these reminiscences, I wish Walter a happy birthday and many happy returns.

**About the Author**: Lu J. Sham is a condensed matter theorist, currently a professor of physics at the University of California, San Diego. He was fortunate to have been taught by Harry Jones (Imperial College, London), John Ziman, Volker Heine, Phil Anderson, Jim Phillips, David Thouless, and Neville Mott (at Cambridge), and Walter Kohn (La Jolla). He is having great fun working with a bunch of collaborators on quantum computing and on spintronics. Contact: Department of Physics, University of California at San Diego, 9500 Gilman Drive, La Jolla, CA 92093-0319, U.S.A.; lsham@ucsd.edu; physics.ucsd.edu/~ljssst/ljs.html

# A Mean Martini

David Sherrington

University of Oxford, Oxford, U.K.

For my wife Margaret and myself, going to La Jolla in 1967 was like going to heaven. It seemed like the Garden of Eden before the fall, with exotic flowers and trees, warm blue sky, golden sands, and a crystal-clear ocean with white crashing waves and coloured fish, very different from the damp, dark industrial Northern England we had left behind. We had even reached it via an appropriately epic journey, seven days on the Atlantic Ocean in a ship designed for the Mediterranean, with inadequate stabilisers, entering the USA for the first time at a bustling New York apparently inhabited by selfish hostile people; followed by a three-day train journey across Canada, through forests, past lakes, across prairies, and then through mountains and gorges; and finally on a Greyhound bus, direct from Vancouver to San Diego. We were tired but the people were friendly and the welcome warm. We had no hesitation.

Above this paradise on the coast was the temple on the hill, the University of California San Diego, home to an impressive panoply of academic "gods" and their students and post-doctoral "acolytes" from all around the world. And I was going to work there. It was a marvellous experience, working with Walter Kohn, learning from him and others, and at the end of each day driving home with a view of a great red globe rapidly descending into a deep blue ocean, to our home 200 yards from the shore, along which we would walk and listen to the waves, watch the sea ebb and flow in the pools at what we privately called "Moon Craters", and pass the seals on our way to our turnaround at the beautiful Cove.

I used to joke, proudly, that my corridor at UCSD had only four Nobel Prizes (one each for Harold Urey and Maria Goeppert-Mayer and two for Linus Pauling). But another was in the making, for that was just after Walter had discovered Density Functional Theory. DFT was one of the topics which he gave to his new post-docs; it was taken up by my contemporary and now-friend Norton Lang. But it was not Walter's only interesting and potentially important recent discovery. I started to work with him on another, excitonic insulators. We explored the possibilities of excitonic phases in symmetry-induced zero-gap semiconductors (such as grey tin, which Walter even agreed to spell in its English way), realised that these materials would have novel dielectric behaviour even without an excitonic insulator transition and investigated it, and went on to consider more general fundamentals of Bose condensation and superfluidity in Fermi systems. In the process I came to appreciate and admire greatly Walter's amazing insights, the clarity of his thinking, his incisiveness, and the elegance of his explanations, both verbal and literary. Of course I expected some of this from his earlier papers, such as his defining series on semiconductor theory with Quin Luttinger and his beautiful paper on the theory of the insulating state. But there is nothing like the real thing.

Walter is always careful to think clearly before he speaks, often pausing for what seems a long time before a well-measured comment. I have tried to learn to do the same, but lack his innate patience. Often were the times in our discussions when I would have to contain myself not to blurt out and interrupt his thoughts. One of my defining memories is of Walter discussing with a student who had learned this lesson, the two sitting in Walter's office with the door open, as usual; one would say something, then a long pause and eventually the other would speak briefly, another pause and so on.

All who have experienced Walter will have strong memories of his great lectures and of his quiet but incisive questioning, which has brought many a too-brash seminar speaker up short. Another, at least for me, Walter-innovation which I enjoyed, profited from and in turn later instigated elsewhere is his brown-bag lunch discussion meeting; we talked about ideas in progress, someone called upon or volunteering to talk without (overt) preparation or even certainty of being correct, testing ideas on the others who responded in an atmosphere of mutual assistance and trust. When I left La Jolla to take up a Lectureship at Imperial College and try to revitalise Theoretical Solid State Physics in the Physics Department, one of my first actions was to set up an analogue of these lunches with the corresponding group in Mathematics. Indeed it was at one of these (Tuesday) lunches, held in a beautiful old room with ebony marquetry and a (possibly apocryphal) history of illicit liaison such as might be the subject of a BBC historical drama, in what is now part of the Victoria and Albert Museum, that I first reported the spin glass model which now bears my name and that of Scott Kirkpatrick (published in my second SK PRL, the first having been with Walter on the frequency-dependent dielectric function of a symmetry-induced zero-gap semiconductor).

Another of Walter's choices, which again I liked so much that I had to emulate it, came later, when he was at the Institute for Theoretical Physics, as its first Director, a job he did so well that ITP has become a magnet for the best students, postdocs and visitors, and a model for many clones. When I first saw it I fell in love with the blue carpet and admired the complementation of the leather seminar room chairs. So in my study in my previous house I had to have a similar smoky-blue coloured carpet and also a complementing leather chair. When I came to Oxford and was allocated funds to refurbish my office the carpet again had to be that smoky blue. (I do not know if it was some invisible influence of Walter or just chance coincidence, but when our son Andrew was born in San Diego my fellow post-doc and friend Norton Lang presented him with a not-dissimilarly coloured blue blanket that remained his dearest possession for many years.)

We have many enjoyable social memories, in many places including Walter's and our homes, but let me mention particularly one in France in the late seventies when I was spending a couple of years at the Institute Laue Langevin. Walter was visiting so we put on a dinner party, with also as guests Philippe Nozieres and Benoy Chakraverty. Margaret prepared a nice meal and I looked out for my best wines. We offered drinks beforehand and Walter volunteered that he mixed a mean martini. So we let him loose. First he put our best glasses in the freezer, much to Margaret's concern but they survived, as he said they would. Then he mixed one of the best and largest jugs of martini I have ever seen, using most of our duty-free gin but just the merest whiff of our white vermouth. They were very good martinis and we all enjoyed them, although I am not sure any of us could properly appreciate the subtleties of the wine thereafter. Everyone seemed to enjoy the evening. When Walter and Mara came to dinner with us at home in Oxford more recently we limited the martinis in favour of giving the wine a better chance. For his eightieth birthday, though, we shall toast him in champagne.

We last met in Zurich just over a year ago, when Walter was on one of his twice-yearly visits. I went specifically to explore further with him the problem of Bose condensation in Fermi systems, which was the subject of a Rev. Mod. Phys. article we wrote back in 1970 but which seemed to have been considered too elegant by several experimentalists. He was in great fettle, the same clear careful thinker I had enjoyed as a post-doc and a great companion as we walked together in the woods behind ETH. We did not have time to do all we wished in the two days I could stay, but I look forward to two months at UCSB later this year, when, as well as Margaret and myself enjoying again the company of Walter and Mara, I hope Walter and I can go a bit further together scientifically.

**About the Author**: David Sherrington is Wykeham Professor of Physics and Head of Theoretical Physics at Oxford. He studied with Sam Edwards (functional integral methods in many body theory) and was a post-doc with Walter Kohn in La Jolla in 1967–69 (excitonic insulators, zero-gap semiconductors and Bose condensation in Fermi systems). He was 20 years at Imperial College and has spent leaves at IBM, ILL, Schlumberger-Doll, LANL and IAS. His current research interests are in the statistical physics of complex, disordered and frustrated systems (in many contexts), in which he has coordinated several European programmes. Contact: Theoretical Physics, 1 Keble Rd., Oxford OX1 3NP, U.K.; sherr@thphys.ox.ac.uk

# Lunch with Walter

Mark E. Sherwin

University of California, Santa Barbara, U.S.A.

Walter is famous for many things, not least of which are his lunches. I have experienced two kinds of lunches with Walter. Weekly "brown bag" lunches in the Physics Department conference room, and lunch parties on his sunny patio. Through these lunches, Walter has nourished me and my family in many ways.

Soon after I arrived at UCSB, as a young assistant professor straight out of graduate school, I went to Walter's office to discuss some of the research problems I was interested in. I was beginning experiments on electrons in semiconductor "quantum wells" driven by intense oscillating fields at Terahertz frequencies. The dissipationless classical dynamics of a single electron in such wells is chaotic, and I was interested in the quantum manifestations of this classical chaos. Walter encouraged me to discuss my ideas at one of his weekly lunches.

At these weekly lunches, I quickly experienced Walter's insistence on precision and rigor in stating and discussing problems. One doesn't simply come to Walter's brown-bag lunch with a prepared talk, give it, and then answer a few questions. In fact, I have never seen a speaker complete a talk in a single lunch. In the first few minutes, Walter will ask for a clarification of a definition or an assumption, and the ensuing discussion may end up consuming the whole lunch hour. The speaker will be invited to continue the following week. Usually, a topic that is initiated by a speaker will continue to be the subject of discussion for several weeks. In this hectic age of information overload, Walter's weekly lunches were an island of contemplative scholarship where truth, precision and rigor were not subordinate to the pressures of time.

Of course, I learned a tremendous amount of physics from Walter and his associates. It turned out that we needed to use time-dependent density functional theory to understand the dynamics of the electrons in our quantum structures. There was no better place on earth to learn about this than in Santa Barbara, from Walter and his associates. But it was also a thrill for me to be able to teach Walter some things, and to pique his interest with the topic of quantum systems driven by oscillating fields of arbitrary strength.

Walter went on to publish two very beautiful papers on this subject. One is a characteristically careful and rigorous look at "Time-dependent Floquet theory and the absence of an adiabatic limit" (Phys. Rev. A **56**, 4045 (1997)). The other is "Periodic Thermodynamics" (J. Stat. Phys. **103**, 417 (2001)). The latter develops a thermodynamics of quantum systems driven by strong periodic fields and interacting weakly with a heat bath. This is a topic of great personal interest, relevant to ongoing experiments in my laboratory, and important for future quantum devices which are driven by strong periodic fields.

I have a lot more to learn from Walter, and one particular area is in the art of entertaining. Occasionally, especially when Mara was spending a lot of time away caring for her father in the 1990s, my wife Cathy or I have gotten a phone call late in the week in which Walter would invite us to lunch at his home that week-end. One of the first times I remember going was when my son Evan, now thirteen, was an infant. Walter always did all of the preparing and cooking – he is particularly good at grilling large pieces of fish. Evan was soon joined by his younger brother Stuart, now 11. Walter and Mara guided Evan and Stuart through their enchanted garden, helped them pick fruit when they were too young to do it by themselves, and opened up a long-dormant sandbox. As far as we know, Evan and Stuart are the only people who have eaten any of the particularly sour oranges that grow in Walter and Mara's garden.

We did share other kinds of occasions with Walter. We were all, of course, VERY EXCITED when Walter won the Nobel Prize. Cathy, Evan and Stuart wrote the appended poem for the ITP's first celebration of this event. Cathy recited the poem, accompanied by Stuart and Evan on party poppers.

Walter, thank you for the lunches!

## A Poem To Walter Kohn

### By Evan and Stuart Sherwin and their Mom
### October 13, 1998

Our friend, Walter Kohn, won the Nobel Prize!
It's the best news we ever heard!

We think we should throw a huge party to celebrate,
Inviting all the living people who ever won the Nobel Prize before.

I think we should get some of Walter Kohn's really, really sour oranges,
We could make some pretty good orange juice that is just like lemonade.

I think we should bring a wedding cake!
Chocolate with chocolate frosting,
With gingerbread men and brownies and hot fudge and chocolate sprinkles
on top,
And the gingerbread men would come to life and write:
**CONGRATULATIONS WALTER KOHN!**,
And slide down the cake right through the hot fudge and on to the table.

We should bring the things that when you blow on 'em a long thing
comes out like a monster's tongue and it makes a party noise like a horn
out of tune,

We should throw confetti – that part will be a surprise!
And we must bring those things that you pull the string back and
it has a tiny bit of gunpowder inside that explodes and shoots
out confetti that has little faces that fly around and say
**CONGRATULATIONS!**

**About the Author**: Mark Sherwin is a Professor of Physics at the University of California at Santa Barbara, as well as the director of the Institute for Quantum Engineering, Science and Technology (iQUEST). His research interests include quantum information processing, the dynamics of quantum systems in strong periodic fields, nonlinear dynamics, optical communications, and sensitive detectors of electromagnetic radiation in the Terahertz ($10^{12}$ Hz) frequency range. These interests are pursued by studying the dynamics of semiconductor nanostructures driven by strong Terahertz-frequency electromagnetic fields. Contact: Physics Department, University of California at Santa Barbara, Santa Barbara, CA 93106, U.S.A.; sherwin@physics.ucsb.edu

# Act of Compassion

Bonnie Scott Sivers

Santa Barbara, U.S.A.

"Bonnie, would you call the Campus Police?" his quiet voice asked. Walter had come through the door connecting our offices, closing it gently behind him.

I was startled by his solemn request, wondering what was wrong. We were in the new stages of the Institute for Theoretical Physics at Santa Barbara and I was Walter's administrative assistant. What could have happened?

"I found a young man wandering in the hall downstairs," he said. "He's wearing only his pajama bottoms and seems confused and frightened. I talked him into the elevator and to coming up here with me. I have him in my office now. Please explain the situation to the Police and ask them to come without fuss. I don't want to upset him any more than necessary."

It was late in the day and Walter was returning from a meeting when he came upon this unusual situation. My first thought was that Walter could be in some danger.

"He seems to be harmless, only frightened," Walter said, perhaps guessing what I was thinking. "He may have strayed from Devereaux." (Devereaux is a facility close to the University for developmentally handicapped individuals.)

I called the Police and explained what had occurred, urging them not to upset this young man when they came. The Police arrived quickly, talked to him for a few minutes, and then kindly led him away. The fellow cried as he left Walter's office; I'm sure he did not want to leave Walter's warm support. When this episode was over, I wondered how many of us would have acted out of such genuine concern in a similar situation.

This is only one of Walter's many acts of compassion that I was privileged to witness while he was the director of the ITP. Others cannot be written about as they would betray confidentiality. His empathy and desire to understand the human condition was not limited to a wandering lad. He gave that same thoughtfulness to everyone: colleagues, visiting physicists, and support staff.

Walter, if I have not remembered your exact words, please forgive me. The act of compassion is, however, the same. You have my utmost admiration.

**About the Author**: Bonnie Sivers was the first administrative assistant/management services officer of the Institute for Theoretical Physics in Santa Barbara, starting at its opening in September of 1979. She previously worked for groups of theoretical and experimental physicists at the University of California in Berkeley. Upon retirement from the ITP in 1993, she began writing her memoir of growing up on Wyoming ranches. In addition, she writes fiction, poetry, and serious, researched environmental articles on the problems of coal-bed methane drilling in Wyoming. How land is used has been a life-long passion. Contact: 1055 Cheltenham Road, Santa Barbara, CA 93105, U.S.A.

# Some Recollections of Life with Walter Kohn

John R. Smith

Delphi Research Labs, Shelby Township, U.S.A.

In 1968, I first met Walter at Case-Western Reserve University, where he was giving a series of lectures. I had completed computations on metal surfaces with the new density functional theory he had published with Hohenberg, and wanted to show the manuscript to him. Walter gave me an appointment, and we had an enjoyable hour discussing the computations. Walter said he had a post doc named Norton Lang who was also interested in carrying out surface computations.

In 1970, I began postdoctoral work with Walter in La Jolla. Walter was one of the first hired to form the Physics Dept. there. He told me that when he visited there and saw whales playing in San Diego Harbor, he knew it was the right place to come to. It was a wonderful time. Walter was a delight to work with because he has such clear insights and understanding of physics and because it was just plain fun. We developed a linear response function for metal surfaces that we applied to chemisorption and other surface properties. A collaboration arose with See Chen Ying (another post doc at the time but now a long-time professor at Brown University), and the three of us published 4 papers. The three of us continued to work together for several years after See Chen went to Brown and I went to General Motors Research.

I know at that time Walter was worried about his heart condition. He had to give up such adventures as hang gliding, which he had longed to do. About 30 years later, I was chatting with Walter at a March APS Meeting in Atlanta. We were walking to the conference center, which was about a mile away. Walter asked if I minded if we were to jog to the conference center, as his post doc was scheduled for a talk and he didn't want to miss it. As I was huffing and puffing to keep up with him, I couldn't help but marvel aloud how he had conquered his condition.

While I was a post doc with Walter, my thesis adviser Jan Korringa came for a visit. On this occasion, my wife and I had Walter, Jan, and their spouses to dinner at our modest apt. Of course I was well aware of what had become known as the KKR theory. Prof. Korringa had always said that he was pleased that Kohn and Rostoker had published their paper, because Korringa's paper had not received much attention. To my surprise, Korringa and Kohn had never met, despite their sharing the authorship of such a well-used band structure method. That's an evening I'll always remember.

Other occasions I'll always remember were Walter's "brown-bag" lunches at La Jolla. Walter strongly believed that such weekly gatherings were valuable and I concurred (I organized such sessions at General Motors). At these lunches, the grad. students and post docs would gather, along with several other professors including Prof. Shelly Schultz, an experimentalist who made many important comments keeping the theorist's feet on the ground. Walter believed these meetings should be impromptu. So the grad. students and post docs would nervously munch on their sandwiches, waiting for Walter to decide on whom he would call to give an update on his or her research to the 30 or so attendees. The pressure was on, but usually valuable comments and suggestions resulted for the speaker.

One story I remember that Walter told me from his younger days was memorable. As Hitler was expanding his power, Walter immigrated to England from Austria. Despite the fact of his Jewish heritage, since he spoke German the British put him in an internment camp. But Walter was, I believe, about 19 still maintaining a strong appetite. The British guards were auctioning the prisoner's bread, and Walter didn't have enough to eat. When the British asked for volunteers to transfer to Canadian imprisonment, Walter quickly raised his hand. Not only was he better fed in Canada, they put him to work folding parachutes – a most responsible position for a prisoner. Perhaps most importantly from the standpoint of his future career, Walter saved his pennies while a prisoner and bought "Modern Theory of Solids", by Seitz. How many prisoners would have done that? Much later, his thesis at Harvard was done with Schwinger, a high energy physicist. It turned out that the job that was available to him at Carnegie Mellon was teaching solid state physics, so his early book purchase paid off. As they say, the rest is history. He spent the rest of his career in solid state physics. Of course his Nobel Prize is in chemistry – more on that later.

Several years later, Walter took on another major responsibility – the launching of the Institute for Theoretical Physics at Santa Barbara. At that time, it was considered to be a pilot program, but under Walter's guidance it developed into permanent facility. I know that Walter put his heart and soul into it. In 1985, Walter hosted a project there on Surface Physics, and I participated there for 6 months. That 6 months was so wonderful that it changed my career. I was headed into major management at GM, and because of the full time research experience with Walter and the rest of the Surface Physics team, when I returned to GM I went back into research full time. It was during that time that Jim Rose and I were able to expand our interests in the universal binding energy relation, work that continued for years after that.

I remember that Walter had a relatively extensive garden in his home with Mara. He told me one time (I don't think he remembers this), that he planned upon his retirement to catalogue all the plants and flowers. This would be accompanied by bus trips through Europe with other retirees. His Nobel Prize usurped all this. I can't tell you how happy I was to learn that his work had been so honored. Walter visited recently, giving a talk at Wayne State and showing clips from the Nobel Ceremony. His density functional theory has finally found the recognition it deserves – from chemists! Walter, The Chemistry Nobel Laureate, enjoys mentioning that he was 16 when he had his last chemistry course. But chemists recognized the value of DFT.

Walter, you have not only been a wonderful scientist but also a good friend. Happy 80$^{\text{th}}$ birthday.

**About the Author**: John Smith is a Research Fellow and head of a group making metal and ceramic coatings as well as computer simulations of machining processes. His current research involves first principles computations at the atomistic level for solid surfaces and interfaces, especially metal/ceramic interfaces. Recently, he is particularly interested in the effects of interfacial impurities on adhesion. The interfaces he simulates are fabricated by a new coating process he is developing called kinetic spray. Contact: Delphi Research Labs, 51786 Shelby Parkway, Shelby Township, MI 48315, U.S.A.; john.r1.smith@delphiauto.com

# Bonjour Mon Très Cher Ami et Bonne Anniversaire

Charles Sommers

C.N.R.S., Orsay, France

"C'est avec beaucoup de joie que je vais parler de nos moments passés ensemble depuis plus que 35 ans. Mais je continuerai en anglais pour que tout le monde peut partager mon bonheur."

As I started thinking about our numerous times together, my mind went from one experience to another. Each thought triggered yet another, reminding me of Joyce's novel Ulysses, and this prompted me to write just as the thoughts enter my mind, in a stream of consciousness way. So here it goes.

I think we first became close friends in 1968 when I was a postdoc with Ben Segall at Case Western Reserve University. Ben invited you to give a series of lectures on exciton theory and I was chosen to take notes and write them up later for those who attended your lectures. This was my first direct insight into how the Kohn mind functioned. In the end, you approved of my write up which wasn't surprising as I asked you midway through if I could record them live to avoid finger cramps. Many times you spoke faster than I could write. The dinners and time spent together that week were what started our long *amitié*. I had done my thesis using KKR and local density theory so I knew who you were, but not what kind of person you were.

In 1970 Carl Moser of CECAM invited me to work in France. I was later able to convince him of the importance of solid state physics in general and band theory in particular which enabled me to invite you to France many times over several years so you could share your ideas and experiences with us. This also provided many opportunities to spend more "off" time together. On one occasion we went to a Vietnamese restaurant in the Latin Quarter where we were surprised to find a pseudo-multilingual menu. We had a good laugh to find *Boulette de Chef* translated as *"small pork ball of the chief"*. Coming to France also allowed you to visit your uncle in Lyon of whom you always spoke so highly.

I remember the difficult period of your divorce and how I convinced you as well as my friend Mel Grossgold, who was in a similar situation to help lay glass tiles for the oversized bathtub at 16 rue du Sommerard as a therapeutic cure.

I remember a peculiar event at the conference in Menton in 1971. John Slater was invited and we were talking together when he came into the center and walked right past us without acknowledging your presence. I think you chalked it up to his jet lag. Later, in a heated discussion on the merits of the Green theory approach to multiple scattering, he stepped off the podium refusing to continue because we couldn't speak to him in a language he understood.

Then there were your apartments: on rue des Ecoles, at my friend Paule Cahn's place opposite Parc Montsouris, the apartment at the Ecole Normale, chez les Sommers, a most memorable one on rue Pierre Nicol which belonged to the American Academy of Paris. The director, Mr. Prince, was to bear the Kohn's frustration since the apartment lacked many basic amenities such as towels and hot water. Our dealings with Mr. Prince reminded me of S.J. Perlman's stories about his stay in Paris and the cleaner's mix up of his laundry with the Pandit Nehru's. He got the Pandit's toga instead of his own underwear.

I remember the visits to my country house near Montargis in the 80s. I'll never forget your description of our "semi-private" toilet (it was separated from the hallway by a blanket suspended from a rod). Speaking of country houses, there is that memorable occasion when you and Mara were coming to visit us in the Côte d'Or; you missed the TGV and wound up taking a local train from Paris going in the same direction. The trip to Montbard took 11 hours instead of 1 hour a record that holds to this day.

Talking about travels, I remember a visit in San Jose when you and Mara were just married and you insisted on taking me to Tijuana as part of a Spanish cultural tour. You both accompanied me to the airport and told me to keep an eye on my baggage on the plane from Santa Barbara to Los Angeles because it sat 20 passengers and they bumped baggage to make room for more people. Sure enough they lost my suitcases!

I remember spending a month in Santa Barbara with Maurice Kléman at a liquid crystal workshop. We cooked many meals together in particular the one to which our friends Shlomo Alexander and John Bardeen were invited. We also spent two days trying to find a way to hang 2 hi-fi speakers that Mara would approve of. Thinking about that house also reminds me of the time just before the renovations when you asked me if I knew of a competent architect. Mel recommended his brother in Los Angeles whom you contacted. His solution was equivalent to rebuilding one side of the hill to support the library, a solution like those we reject for many of our problems in physics.

Of course I remember our experiences with our dear friend André Blandin. During one of his difficult periods you invited him to spend some time at the ITP in Santa Barbara. There he met a French woman (whom I later met in Paris at his funeral). She told me the following story which was typical of André and why we loved him so dearly (he was such a special person). They had been invited to dinner by one of your notable colleagues. Sitting on André's plate was a mound of cauliflower in a cream sauce, a food he detested since his childhood. When the husband was called into the kitchen by his wife to cut the meat André got up to go to the toilet, taking the cauliflower with him. About 20 minutes later he came back and as he was sitting down his girlfriend noticed an enormous white stain coming from his pant's pocket. Later she asked him what took him so long in getting back to the table. He calmly explained that he had tried to flush the cauliflower down the toilet in several pieces but that it got stuck and wouldn't evacuate. Finally, realizing that people were probably waiting for him, he just put the rest in his pocket and came back to the table.

Another memorable meal was the traditional Passover seder we had with you, Mara, Mel, and my cousins Lee and Len Herzenberg from Stanford. You were of course chosen to be one of the wise men. At the end I even took out my trumpet and accompanied the singing of the Passover songs.

I remember when you became my *sous-chef* (we still make a good team in the kitchen) in Paris and we started our traditional dinners for *"les fidèles"* (Balian, des Cloiseaux, Jérome, Sommers, Savin) which are repeated every time you visit. Your bypass surgery has made you *un jeune garcon* and you haven't stopped traveling since. I recall the touching speech you gave at Jacques Friedel's 80th birthday celebration at l'Académie des Sciences last year. It showed how beautifully you manage to turn professional relationships into deeply personal ones.

I could probably continue for several pages, but there's work to do.

"Alors, au boulot mon sous-chef *Kohn Du Sommerard* les fidèles arriverons bientô                                                                     e sur du riz brésilien accompagné d'un Volnay Les Santenots 1996 ou bien un gigot piqué d'ail au four avec des feuillettes de pomme de terre aux échalotes, suivie des fromages époisses et délices de bourgogne à consumer avec un Meursault Charmes 1998. On n'est pas encore sorti de l'auberge avec tant de décisions à prendre."


**About the Author**: Charles Sommers is director of research at the C.N.R.S. in France. He works at the Laboratoire de Physique des Solides at Orsay. His research concerns condensed-matter theory specializing in magnetic multilayers. His current interests include developing and adapting the fully relativistic spin polarized screened KKR method (using density-functional theory) and the Kubo formalism in order to study transport and optical properties of nanosystems. Contact: Université de Paris Sud, Bat. 510, campus d'Orsay, F-91405 Orsay, France; sommers@lps.u-psud.fr

# Walter Kohn and Boris Regal: The Early Days of ITP

Robert Sugar

University of California, Santa Barbara, U.S.A.

Walter's service as the Founding Director of the Institute for Theoretical Physics was certainly one of his major contributions to theoretical physics. I had the good fortune to serve as the Deputy Director during the first two years of his term, and again during the last year. As a result, I was able to watch at first hand as Walter turned ideas of Jim Hartle, Boris Kayser, Ray Sawyer, Doug Scalapino and myself into a vibrant research institute, which quickly became a leading center for theoretical physics. This was one of the most exciting experiences of my professional career, and I have always been grateful to Walter for making it possible for me to participate in it. It is well known that the original idea for a National Science Foundation funded Institute for Theoretical Physics was due to Boris Kayser, the long-time Program Director for Theoretical Physics at the NSF. Boris Kayser was a strong supporter of ITP, and Walter formed a close working relationship with him. It is perhaps less well known outside of Santa Barbaria that Walter was also in constant contact with another high official at the NSF, Boris Regal. Regal carried the title Theoretical Director of Experimental Programs.[1] Whereas Boris Kayser focused on solving problems, Boris Regal focused on creating them. Naturally, Walter and Regal bonded immediately. The definitive history of the early days of ITP is contained in the treatise *A Brief History of the Origins of the Institute for Theoretical Physics Including a Description of How Walter Became Director*, which Regal presented to Walter on his seventy-fifth birthday. Here I only have space to set out a few highlights of Walter's tenure as director. One of Walter's first objects was to establish close relations with officials of the NSF. The extent of his success can be gauged by the following telegrams which he received from Boris Regal. The first arrived on the opening day of the Institute, September 1, 1979. It was brief and to the point:

---

[1] Regal's direction of programs was entirely theoretical, and ITP was certainly an experiment.

Dear Dr. Kohn:

Congratulations on the successful opening of the Institute.

Boris Regal

The second arrived on September 2, 1979. It reads as follows:

Dear Dr. Kohn:

In view of the successful start of your Institute, the Physics Division has decided that it can easily sustain a 50% budget cut. This action indicates the confidence the NSF has in you and the UCSB Administration.

Boris Regal

This news was not as bad as it might appear. At the time, the Institute's budget from the NSF was $0.00. In fact the University did not even have a letter indicating the NSF intended to fund the Institute. It is true that there had been a phone call from Regal the previous spring in which he indicated that the NSF would support the Institute. Apparently Regal had meant moral, rather than financial support. By the end of November Walter became concerned, to say nothing of the UCSB Administration, which was paying the bills. As a result, one can imagine the extent of Walter's relief upon receiving the following letter from Boris Regal.

December 2, 1979

Dear Dr. Kohn:

I am very sympathetic to your request for a letter indicating that the Foundation intends to support the Institute. One will be forthcoming shortly. (No later than August 31, 1984.) As for your request that the Foundation also provide financial support for the Institute, I can only say that it will be given serious consideration; however, I should advice you that the Foundation is very reluctant to risk disrupting the funding pattern of a successful operation.

Sincerely yours,
Boris Regal

August 31, 1984 was, of course, the last day of the original ITP grant. One of the problems Walter encountered when ITP opened was that it had no computing facilities of its own. Members had to use computers located elsewhere on the Santa Barbara campus, which they accessed via dumb terminals located in a single room in the Institute. Walter insisted that terminals not be placed in the offices of Institute members for fear the over-use of computers would destroy their thought processes and distract them from doing real physics. Walter was, of course, an expert on this danger, since, as the inventor of density functional theory, he was then, and remains today, responsible for the burning of more computer cycles than any other person who has walked the earth. Although I recognized Walter's superior wisdom in this area, I was nevertheless concerned when I learned that a very distinguished computational physicist would be a member of the visiting committee for the first external review of ITP. My fear that the Institute would receive poor grades for its computing facilities proved unfounded, as Walter had predicted. However, a short time later, on Walter's birthday to be precise, a gift arrived for the Institute. The note from the anonymous donor indicated confidence that the gift would improved ITP's computing capabilities by several orders of magnitude. He/she was certainly correct, as the gift was a beautiful, state of the art abacus. Walter jumped to the conclusion that the gift was from his friend Boris Regal, but a close inspection of the postmark on the package ruled out that possibility. The identity of the donor remains a mystery to this day. Walter gave the abacus a place of honor on a shelf immediately behind his desk, thereby protecting Institute members from its nefarious effects, at consider risk to his own well being. By an amazing coincidence, the ITP abacus disappeared on the very day that Walter's term as director ended. Although it is true that Walter moved to an office on the physics building on that day, there could be no question of his having taken the abacus with him, as it had clearly been presented as a gift to ITP. It was not until sometime later that I noticed that the abacus had migrated to the physics building on its own accord. In order to protect members of the UCSB Physics Department and of ITP, Walter had confined the abacus to the top of his desk. Just how nefarious its effects could be became apparent shortly, as not one, but two silicon based computers soon joined the abacus on Walter's desktop. Although the constant use of these machines has had no noticeable effect on Walter, exposure to computers at this level is not recommended to anyone less strong minded. Surprising as it may seem today, ITP was quite controversial when it begin operation. Initial support was in the form of a five year grant, and it was far from certain that the grant would be renewed despite the Institute's great success. In another of those amazing coincidences that always seem to follow Walter's activities, the National Science Board was scheduled to take up the question of the Institute's renewal on the same day that a celebration of Walter's sixtieth birthday had been scheduled. In keeping with Walter's tastes, a small meeting and a large banquet had been planned. There was considerable concern that the celebration would turn into a wake, but the tension was relieved when the following telegram arrived from Boris Regal:

March 7, 1983

Dear Dr. Kohn:

On behalf of the members of the Physics Division of the National Science Foundation, I would like to wish you a happy sixtieth birthday. Unfortunately, I am unable to do so. As you know the NSF cannot take actions of this import without prior approval from the National Science Board. However, I am happy to be able to inform you that the Science Board has agreed to put this matter on its agenda for next week's meeting. At the same time it will consider the Physics

Division's requests to send you felicitations on your fifty-ninth, fifty-eighth and fifty-seventh birthdays. Understandably, with such a heavy docket, the Science Board has had to postpone consideration of the Institute's renewal. Rest assured that the renewal question will be rescheduled for the next opening on the Board's agenda, possibly as early as its April meeting in 1987. I trust this good news will contribute to the festive mood appropriate to your birthday celebration.

Sincerely yours,
Boris Regal

As was his custom, Regal sent the message to me with a request that I read it at the banquet. Unfortunately, a very high NSF official who happened to be in attendance confused the reader (me) with the author (Regal), and severely berated me for being disrespectful to the NSF. Not wanting to take credit for another person's work, I set the official straight. This may have been a mistake. Shortly thereafter Regal was forced to resign from the NSF, and we did not hear from him for many years. I recently discovered that Boris Regal has taken a position in the White House with the assignment of filling any holes that might exist in the President's knowledge of science or science policy. When I contacted him to inquire whether he would be willing to contribute to this volume, he informed me that he was far too busy, as his new position forces him to literally work around the clock. However, he did ask me to include the following message to his old friend in my own contribution.

Dear Walter,

I would like to wish you a happy birthday on behalf of the entire administration, and I have requested permission to do so from its highest authority. I am confident that Dick Cheney will give me his decision the next time he emerges from his bunker. (The contrast between the decision making process in the White House and the NSF is quite remarkable). I remember working with you during your days as Director of ITP with great pleasure, and I have been delighted to see that many of the theoretical ideas that came out of the Institute under your leadership are having a practical impact on Washington under the President's leadership. Let me cite a few examples. The very first paper you wrote as Director of the ITP, indeed the very first ITP preprint, *Experiments on a Certain Class of Fermented Liquids*, has become the K Street bible. Now that Congress has repealed its outlandish restrictions on lobbyists' contributions to its members, I predict that your important discoveries will play an even greater role in lubricating the wheels of government. I am also pleased to be able to inform you that the famous ITP Icing Machine has become the administration's ultimate weapon in the fight against global warming. We are convinced that it will win the battle single handed, eliminating the need for burdensome regulations on the emission of greenhouse gases or for any slowdown in the growth of SUVs. I know how proud you were Dr. Wilczek's pioneering work on the invisible axion while you were director, so I am sure that you will be even prouder to learn that the administration plans to base the next generation of missile defense on invisible axion beam technology. Although it is true that this technology will only work if our adversaries equip their missiles with invisible axion attractors, who can doubt that the deft diplomats in our administration will be able to convince them to do so. As I am sure that you have noticed, the budgetary strategies I devised for the ITP are now being applied to the country as the whole. Just think of the state governments as being equivalent to the UCSB administration, and it will all fall into place. There is one favor I would like to ask in this regard. I am sure that you are aware that the computing facilities of some of the key agencies involved in homeland security are significantly behind the state of the art. Since recent and planned tax cuts will prevent the government from buying new computers for the next few decades, I hope that you will be willing to loan these agencies the famous ITP abacus. I am certain that it will improve their computing capabilities by several orders of magnitude. I shall be sorry to miss the festivities surrounding your birthday. I hope you will offer a toast, or two, or three, or four on my behalf.

Your friend,
Boris Regal

**About the Author**: Robert Sugar is a professor of physics at the University of California, Santa Barbara. His main scientific interests are in the study of quantum chromodynamics and strongly correlated electron systems. His research in both of these areas involves large scale numerical simulations. He served as Deputy Director of the Institute of Theoretical Physics for three years during Walter Kohn's term as Director. Contact: Department of Physics, University of California, Santa Barbara, CA 93106, U.S.A.; sugar@physics.ucsb.edu

# Flashback to My Post-doc Days with Walter

Yasutami Takada

Institute for Solid State Physics, University of Tokyo, Tokyo, Japan

On the day next to Thanksgiving in 2002, Walter and I were discussing some problems in the density functional theory (DFT) in the office assigned to Walter's visitors in the Physics Department of UCSB (Broida 6113). On the blackboard, Walter, knitting his brows, was writing mathematical equations only little by little, indicating clearly that he was asking himself deliberately whether these equations are pertinently describing our physical ideas or not. This Walter's special style of thinking suddenly reminded me of my younger days in Santa Barbara, dating back to the years 1983–1985 when I was a post-doc of Walter and engaged in the study of atom-surface interactions. In those days I spent many hours with Walter by doing exactly the same way in the director's office of the ITP, his office in the Physics Department (Broida 6111), and even at his home on the weekends.

In my second year in Santa Barbara (and my fourth year in the United States), I was rather depressed; although I published a paper with Walter in the Physical Review Letters in my first year, I could not get any offer for a next job. I, then 34 years old, was a little too young to get an associate professorship in most universities in Japan but a little too old to get an assistant professorship in theoretical physics in major universities in the United States. In this circumstance and thinking of my family to support, I decided to seek a job outside of physics. After a few interviews I actually got an offer from some Japanese companiy to work in its branch in Los Angeles. In the end, however, I did not accept the offer, because Walter suggested to me to hang on to physics by explaining me how highly he acknowledged my talent. At the same time, he promised me to act positively to find a suitable next job for me. After a while, I could get an offer of an associate-professor position from the Institute for Solid State Physics (ISSP), University of Tokyo. I could easily imagine that a recommendation letter from Walter to Professor Yutaka Toyozawa, the director of the ISSP at that time, exerted a decisive power in the process to select a finalist among many strong candidates for the position. Figure 1, taken by Mara, is a snapshot at the party held at the Walter's home about a month before I left for Japan to assume the job.

At the garden of the Kohns' home on 14 April 1985

As far as I know, Walter visited Japan four times; the first visit was very long ago, in 1965, to deliver a lecture on the density functional theory in its very early stage, entitled "A New Formulation of the Inhomogeneous Electron Gas Problem". He was chosen as one of the speakers for the Tokyo Summer Lectures in Theoretical Physics, organized by the late Professor Ryogo Kubo. From time to time Walter told me how much he was impressed with the discussions with Professor Kubo. It seemed that they could understand each other well in spite of a large difference in the cultural backgrounds.

Other three visits were relatively recent ones, in 1989, 1990, and 2002. At each visit, Walter came to my Institute on the way to and/or from the primary destination of the trip in order mainly to spend a few days with me. In particular, Walter stayed a rather long time in Tokyo in 1989. During that period, he wanted to know old Japanese cultures as much as and as deep as possible; for example, we went to the Noh Theater. (Incidentally Walter did not want to go to the Kabuki Theater because of its excessive exaggeration, reflecting the fact that Noh is spiritually much deeper than Kabuki.) We also looked around museums like the Nezu Museum, displaying many important national treasures and assets. (I would confess that I did not know the Nezu Museum to be so famous until Walter mentioned the name at that time.) Figure 2 is a photograph taken with my children when we visited the Imperial Palace (the old Edo Castle).

The last visit in 2002 was a trip following a visit to the Mainland China. Immediately after we met at the Narita Airport, Walter began to tell me how much he was impressed with the oriental philosophy due especially to Confucius. By the time he left my Institute, he proposed to me to spend a few

At the Imperial Palace on 3 September 1989

months in Santa Barbara so that we could begin to make some collaborative work on the DFT for either the ground state or its time-dependent version. Of course, I gratefully accepted his proposal, which led to the discussion mentioned at the beginning of this article.

It turns out that this seven-week stay in Santa Barbara is very important for me. In particular, I am overwhelmed by the energy towards physics Walter maintains; in all aspects, he is just the same as he was in my post-doc days. His energy refreshed me so completely that I felt as if I were still young enough to begin with any kind of difficult tasks in physics from now. Once again, I realize that Walter is my superb mentor.

Finally I would also mention that Mara is a good teacher for my wife, Reiko. Reiko is very grateful to Mara for her excellent English class. Thank you, Walter and Mara!

**About the Author**: Yasutami Takada is an associate professor at the Institute for Solid State Physics, University of Tokyo. He serves as a divisional associate editor for the Physical Review Letters since 2001. His research is concerned with the many-body problem at large in solids, addressing electron correlation and/or effects of electron-phonon interactions. Contact: Institute for Solid State Physics, University of Tokyo, 5-1-5 Kashiwanoha, Kashiwa, Chiba 277-8581, Japan; takada@issp.u-tokyo.ac.jp; www.issp.u-tokyo.ac.jp/labs/theory/takada/index-e.html

# A Challenge to My Own Scientific Thoughts

Walter Thirring

Erwin-Schrödinger-Institute, Vienna, Austria

What a delight to be asked to write about a friend and a great scientist of my generation. Walter is going to celebrate his 80th birthday and is as sharp as ever. Still I make my contribution with a smiling and a weeping eye. A smiling eye since we are glad or even proud that somebody who got his cultural background in Vienna contributed so richly to physics and is worldwide recognized. A weeping eye, because he is not the only one in this category, the list of such people, Erwin Schrödinger, Wolfgang Pauli, Kurt Gödel, Viktor Weisskopf, Liese Meitner, Robert Frisch, Eduard Salpeter, Robert Karplus, Martin Deutsch, Wolfgang Rindler, ... is sheer endless. Had we been able to keep all these people here we would be a scientific worldpower. However the madness of the "Nazis" destroyed it all and we were left with a heap of rubble. To re-establish such a rich tradition takes more than a lifetime and we can only try to be on our way.

I met Walter for the first time in the United States with common friends, in particular with Res Jost and immediately felt that we are tuned to the same wavelength. Later we all appreciated that he frequently visited Vienna, in spite of the terrible experience his family made here after the "Anschluss". Unfortunately his sequence of visits was interrupted when he refused to visit his "Waldheimat".

Though I could understand his decision I could not fully agree with it. Not that Waldheim had my sympathy, but the allegation that he was a war criminal could never be proved. He certainly acted stupidly on several occasions, but to boycott all countries where the president had acted stupidly would create severe travel restrictions. More seriously I had witnessed in the "Nazitime" criminal acts which are of such an order of magnitude different from Waldheim's cowardly attempts to conceal his wartime that I felt one has to make a clear distinction between Waldheim and a real war criminal. In the course of time I found that Walter's and my opinions about American policy and the Near East conflict converged, but nevertheless it seems that this opinion is not so widespread to determine the future.

What Walter makes so likable is his unpretentious and pensive behavior. This does not seem to correspond to the image of the successful American and yet Walter was very successful in building up the physics departments in San Diego and in Santa Barbara. Perhaps scientific competence and personal charm are after all more important than great words. I will leave here his biography because usually the rests are only variations of the theme "he is such a nice guy". To do justice to Walter's intellectual standing I should recollect the challenge he posed to my scientific thoughts.

When we got started into research everybody was fascinated by the covariant perturbation theory in Q.E.D. developed by Dyson, Feynman, Schwinger. The main concern was to get rid of the infinties in the terms of $n^{th}$ order, but the question of convergence seemed out of reach. As a first step in this direction Walter, together with Res Jost determined the validity of the Born-approximation in wave mechanics. Their result that things get sour once a bound state develops was to be expected as the scattering length jumps at this instance from -infinity to +infinity. So this was a solid result conforming with common sense, but people working in quantum field theory mostly sneered at elementary wave mechanics. A further result of Walter in this field, the Kohn variational principle in scattering theory, was quite a surprise. One had the feeling that the scattering data are the result of interference of waves and thus an absolute bound seemed out of question. However this feeling is based on too narrow a point of view which looks only on the wave equation. But scattering theory can also be formulated in terms of operators and in this formulation Walter's various principle becomes an operator inequality.

Walter's most popular contribution is the density functional theory. It is a rich field also for the application of functional analysis since it has many surprising aspects. Even entropy theory can be of some use. D. Mermin, one of Walter's postdocs who became famous, generalized Walter's results to finite temperatures and some unevenness is then smoothed out. Ever since the mathematical technology advanced and Mermin's result can be improved. The key fact he used was the positivity of the relative entropy. It turned out that it is not only positive but bounded from below by half the square of the norm of the difference of the density matrices.

These few remarks are meant to illustrate how Walter's independent thoughts proved to be an inspiration and a source of further progress. We wish him all the best for his anniversary!

**About the Author**: Walter Thirring, Honorary President of the Erwin-Schrödinger-Institute (Institute for Mathematical Physics), Vienna; Professor Emeritus for Theoretical Physics, University of Vienna. Contact: Erwin-Schrödinger-Institute, Boltzmanngasse 9, A-1090 Vienna, Austria; walter.thirring@univie.ac.at

# Walter Kohn's Influence on One Engineer

Matthew Tirrell

University of California, Santa Barbara, U.S.A.

My first personal encounter with Walter was a rather distant one, in fact one that Walter was not even directly aware of, but one that was the first step on a path that had a profound and direct effect on the course of my own career. In 1982, the Institute for Theoretical Physics at UCSB organized a major workshop on Polymer Dynamics (I still have the t-shirt). I was five years beyond my Ph.D. that year, a recently tenured associate professor at the University of Minnesota, and deeply involved in research on polymer diffusion in entangled polymer fluids. The ITP workshop included participation and remarks by Walter both on the field of the workshop and on the general nature of ITP.

I remember Walter as not only being engaged in the in the workshop but also being, then as now, regarded with exceptional respect and collegiality, among a very distinguished group of soft condensed matter physicists, peppered with the occasional engineer. I am sure that the openness, indeed proactive inclusiveness, which has characterized ITP's approach to disciplines other than physics from the beginning, is due in no small part to Walter Kohn's ecumenical view of science.

Not only was that workshop an exciting and stimulating boost to my career and the research directions I was pursuing at the time but it was the definite beginning of a series of interactions I had with the University of California, Santa Barbara, culminating seventeen years later and four years ago in my permanently joining the faculty of UCSB. This initial encounter with UCSB, via the ITP led by Walter, was followed three years later by an offer to join the UCSB faculty of chemical engineering. The College of Engineering in 1985 was led by Robert Mehrabian, an incredibly dynamic and persuasive leader, who is arguably the individual most responsible for leading UCSB to preeminence in materials research. Nevertheless, I declined the offer. However, the fine early impressions gained through the experience of considering this opportunity, coupled with my early ITP exposure, created a very strong base of affinity and admiration for UCSB in me.

Twice more, in 1989, and again in 1994, I returned to ITP for a month or so each time to participate in programs on polymer and biomolecular physics. In the intervening years, my own research interests turned more toward surface properties of macromolecules. One of several lines we pursued in this area, in collaboration with Arup Chakraborty (now chair of chemical engineering at U.C. Berkeley) and Ted Davis (now dean at Minnesota), was attempting to understand the chemistry, both macromolecular and electronic structural, of polymer adhesion to solid surfaces. Density functional theory proved to be a powerful tool to give insight into the nature of chemical bonding at polymer-solid interfaces. A series of publications in the late 1980's and early 1990's with this set of collaborators showed, for example, how common polar functionalities, such as carbonyl groups, play the key role in creating chemical bonds at many interfaces of practical importance. This finding is one of many examples of why and how engineers have been enabled to tackle important practical problem via tools that Walter and his work have provided.

This chain of interactions with UCSB, all routed in organizations and scientific innovations led by Walter Kohn, led me steadily and, in retrospect, rather relentlessly to acceptance in 1999 of offer of the position of dean in the College of Engineering at UCSB. The collegiality and interdisciplinarity instilled early in the ITP under Walter's leadership remain a hallmark, in fact, defining characteristic of UCSB's approach to research. All of us at UCSB owe Walter a great debt for his scientific leadership and also for his example as a creative, inspirational colleague.

**About the Author**: Matthew Tirrell received his undergraduate education in Chemical Engineering at Northwestern University and his Ph.D. in 1977 in Polymer Science from the University of Massachusetts. He is currently Dean of the College of Engineering at the University of California, Santa Barbara. From 1977 to 1999 he was on the faculty of Chemical Engineering and Materials Science at the University of Minnesota, where he served as head of the department from 1995 to 1999. His research has been in polymer surface properties including adsorption, adhesion, surface treatment, friction, lubrication and biocompatibilty. He has co-authored about 250 papers and one book and has supervised about 60 Ph.D. students. Professor Tirrell has been a Sloan and a Guggenheim Fellow, a recipient of the Camille and Henry Dreyfus Teacher-Scholar Award and has received the Allan P. Colburn, Charles Stine and the Professional Progress Awards from AIChE. He was elected to the National Academy of Engineering in 1997, became a Fellow of the American Institute of Medical and Biological Engineers in 1998, was elected Fellow of the American Association for the Advancement of Science in 2000 and was named Institute Lecturer for the American Institute of Chemical Engineers in 2001. Contact: Office of the Dean, College of Engineering, University of California, Santa Barbara, CA 93106-5130, U.S.A.; tirrell@engineering.ucsb.edu; www.chemengr.ucsb.edu/people/faculty/tirrell.html

# A Tribute to Walter Kohn

Ulf von Barth

Lund University, Lund, Sweden

My relationship with Walter Kohn has gone through violent oscillations over the years.

My first encounter with the great man was extremely positive from my point of view. I had submitted a manuscript to a conference at Taormina, Sicily in the late summer of 1973. My talk was scheduled in the beginning of the conference and was on the application of local-density approximations to the self-energies and the quasi-particle structure of simple metals. It was to a large extent an application of theories laid down in the well known Sham–Kohn paper from 1966. The talk was one of my first international appearances and I was rather tense. And I certainly did not ease up seeing Prof. Kohn sitting in one of the first rows. I remember very little from the actual talk but somehow felt that it was a complete disaster. I remained in a state of confusion and dispair until the very end of the conference when Prof. Kohn was asked by the organizers to sum up the conference. I do not remember the exact wording but Prof. Kohn said that he had enjoyed most of the very nice and important contributions at the conference but that he had been particularly impressed by the novel contributions by some of the younger participants. And then he spent a large part of his summary on my talk. It is hard to imagine a more encouraging start of a career in physics.

I do not remember having any illusions of grandeur due to my experience from Taormina but whatever feelings of pride and adequacy I harvested at the time, were abruptly taken away not too long after. We had in the fall of 1973 in Lund calculated the relaxation energy of a proton in the electron gas and discovered, as we thought, a discontinuity in the energy when the density of the gas was made low enough for a bound state to appear. This seemed to be a counterexample to the famous Kohn-Majumdar theorem which states the continuity of the density and energy as functions of the strength of the effective potential, albeit for non-interacting electrons. We went through the proof of Kohn and Majumdar (KM) and thought that we had discovered a flaw in their chain of arguments. A preprint was sent to Prof. Kohn who quickly responded by very politely asking us not to publish the manuscript or, in the case we had already sent it off to some journal, Prof. Kohn wanted to know to which one. We reexamined our calculations and found that we had made a trivial but nevertheless devastating error. It was corrected and the energy as function of density of the gas appeared as a perfectly smooth function at all densities. But we still believed there was a flaw in the proof by KM, something that I got the chance to discuss with Prof. Kohn in person when we met at a CECAM workshop in Paris in the summer of 1975. Unfortunately, we were interrupted and never got the chance to carry the arguments to a conclusion. I was, however, not too eager to bring up the issue at some later time because, as it turned out, the arguments were within the realm of pure mathematics, a field in which I felt the background of Prof. Kohn was far better than mine. In the evening of the same day there was a party in the home of Chuck Sommers in Rue de Sommerard in the Latin Quarter of Paris. Prof. Kohn was standing in the middle of a large group of younger physicists who had gathered around the great man. I too went there and heard them discussing the pleasure of swimming in the ocean outside Santa Barbara. I quickly joined in by saying that it was too cool to swim there because there is a cold off-shore stream in the ocean. There were a couple of minutes of complete silence after which Prof. Kohn slowly turned toward me and said "Aren't you ever right about anything?" I was strongly hoping for some magic 'Abra-Cadabra' that would make me instantly invisible.

In 1977 some people including Prof. Kohn and myself had been approached by Prof. March at Oxford and the late Prof. Lundqvist at Chalmers. We were to write a book on the state-of-the-art of density-functional theory (DFT). The work on the book did not progress very rapidly in the beginning. But one summer, I cannot recall the precise year (in the late seventies and early eighties I was visiting IBM Yorktown Heights every summer for some extended period), I met with Walter (I vaguely remember that we on this occasion became Walter and Ulf) at IBM and we had a very pleasant conversation on how to split the work on the book between us and the other contributors. Not too long after that I got a finished manuscript from Walter and his coauthor, Priya Vashishta. Unfortunately, as is known by many colleagues, speedy publication is not one of my virtues and my coauthor Art Williams and I started to receive letters from Walter in which he expressed his understandable concerns about his contribution becoming obsolete. It was clear that our relation was somewhat strained during this period. I met Walter next time at a DFT conference in Alcabideche outside Lisbon in the summer of 1983, and I could then assure Walter that Art's and my manuscript had then been completed for some time. My talk was on the unphysical behavior of the exchange-correlation potential of DFT and I had some fears that Walter might be critical toward these ideas, but he just gave me a big smile.

Since that time in Portugal, my interactions with Walter have been of a pleasant kind. We have met at conferences and we have exchanged notes on different topics within DFT. One of our more funny encounters was when we both had been invited by Prof. Rao to a local conference on DFT at the

Royal Inst. of Technology in Stockholm. This was in the summer of 1996, I believe, and the conference was actually more of a summer school. Prof. Rao is well known as a charming, very helpful, but very persuasive person. By chance, I ran into Walter outside a restaurant not far from my hotel. I was delightfully surprised and asked him if he had a copy of the program because I did not know when my talks were scheduled or who the other speakers were. To my surprise Walter told me that he had asked Prof. Rao the same question when he arrived. Walter had received the somewhat unusual answer that Prof. Rao was working on the program but would actually like some assistance from Walter in setting it up. Well, I guess that they sat down to work on it and then Walter and I lay the last pieces to the puzzle at a nice cafe that same pleasant evening in Stockholm. The weather was beautiful and warm, Stockholm can be a wonderful city in the summer, and Prof. Rao took us in his car to just any place we wanted see. The lectures went smoothly and I hope Walter also left Stockholm a few days later with a very pleasant experience behind us.

I saw Walter most recently at a DFT meeting in El Escorial outside Madrid during which the dreadful events at the World Trade Center stunned the world. Walter gave a very balanced deeply touching talk on the events in which he stressed the truly international character of research and the responsibility we have as scientists to contribute to love and understanding among all human beings.

From a personal point of view, I was delighted to learn that Walter very much appreciated the work my postdoc Stefan Kurth and I had done on response functions within time-dependent DFT. Walter invited Stefan to spend time in Santa Barbara in the summer of 2002.

I congratulate Walter for a long and oustanding career in research and wish him many more productive years ahead. I still think that, compared to the physicists, the chemists have shown better scientific insight and understanding in awarding Walter the Nobel Prize – but who knows what the future might bring.

Happy birthday dear Walter!

**About the Author**: Ulf von Barth is a professor of physics at Lund University in Sweden. He has been working in the area of many-body theory, the spectral properties of solids, self-energies and one-electron excitation energies and the development of static and dynamic density-functional theory. He has been a guest researcher for several years at IBM Research at Yorktown Heights, for one year at Queen's University in Canada, and for half a year at CECAM in Paris. Contact: Solid State Theory, Physics Dept., Lund University, Solvegatan 14, SE-22362 Lund, Sweden; barth@teorfys.lu.se

# Encounters with Walter:
# From the Mysterious Second "K"
# to the Local Secretary

Peter Weinberger

CMS, TU Vienna, Vienna, Austria

## Learning About the 2$^{nd}$ "K"

I first ran across Walter's name in the beginning of the 1970s, when I started doing my post-doctorate in Uppsala and became familiar with the Korringa–Kohn–Rostoker method – which was to accompany my scientific career for the next 30 years. Almost immediately after my arrival in Sweden, the by-now famous "2/3" battle between the Slaterites and the "disrespectful" rest – i.e., Kohn and company – reached the shores of Europe. At that time "W. Kohn" was no more to me than a name on a few publications. I have to admit that this state of ignorance about his relation to Vienna lasted for almost another 20 years before rumors of such connections reached (even) me.

## Meeting Him for the First Time

In 1989 I somehow became aware of the fact that Walter was to meet his old friend Walter Thirring in Vienna, probably on account of a seminar to be given in the Theoretical Physics Institute of the University of Vienna. I summoned whatever courage I had and took a book I had recently published on the KKR method and – more or less uninvited – showed up at Thirring's office. With a "Knödl im Hals,"[1] I handed over to Walter a copy of the book, trying to point out the dedication[2] I had written in it on account of the 50$^{th}$ anniversary of the beginning of World War II. Walter – wearing what was to become to me a quite familiar cardigan – took the book and almost immediately continued his conversation with Thirring. Well, I thought to myself, at least I had met

---

[1] Viennese expression meaning with apprehension. Literally translated: "with a dumpling in the throat".

[2] Dedication: "To all those scientists and scientists to be whose hopes and lives were jeopardized or extinguished by the criminal Nazi regime in Germany and Austria 50 years ago"

him for three minutes. Not really a big success in establishing relations, I told myself, trying to view things realistically.

## Meeting Him Again: The Ice Begins to Break

But we were to meet again, since I had persuaded my own university to award him a honorary doctorate. There was, however, the need to ask him whether he – a refugee from Vienna – would in fact accept a "Dr. h.c." from a Viennese university. It was probably during a March meeting of the APS (American Physical Society) when we finally sat down and started to communicate. He talked very long about his mixed feelings with respect to Vienna, from where he was chased out and had narrowly escaped to England. He talked to me about the wonderful teachers he had in the (Jewish) Chajes Gymnasium, the last school he could attend in Vienna, and – again and again – about the shameless, criminal attitudes of the Viennese population in the period after the *Anschluss*. I still think he was a bit amazed when I finally replied that all his reservations with respect to Austria and Vienna in particular were very well-known to me, since they were shared by parts of my own family whose fate closely followed his. This long conversation we had was indeed the beginning of a long friendship, which over the years has given me the feeling of being a kind of "adopted son" of Walter. Thinking back, it was probably very characteristic that all the time we spoke English; only much later we started to use also German.

By the way, since for the ceremony of his honorary doctorate he – as usual – had brought along only his western tie, I had to lend him a more conventional one. Occasionally when I wear this tie and Walter is around, he keeps asking me whether I wear "his tie." Perhaps I should hand over to him this tie on one of the next times he is honored.

## Stadlau and the Akademisches Gymnasium

One summer Walter came to Vienna *en route* from a trip to Italy and stayed with us in our little summer house in Stadlau, near the *"Alte Donau"*, a kind of lake within the city limits. Most of his family had already left for their respective holiday destinations, so we had plenty of time to enjoy a few days with him just sitting around in the garden, listening to his stories. As usual my father-in-law popped in several times. During their first encounter something incredible happened: they almost immediately found out that they had attended the same school (*Akademisches Gymnasium*) in Vienna, and had even had some of the same teachers, my father-in-law having been some two grades above Walter. Furthermore, they both had been interned on the Isle of Man and subsequently shipped over to Canada as "enemy aliens". For quite some time they exchanged old school stories with enormous pleasure, until finally our neighbour, an old friend of ours, who had to emigrate first to Cuba and only then was granted admission to the United States, came over and started to contribute to their nostalgia. Walter was somewhat amazed: this no longer was the "postwar" Vienna he had had in mind.

## A Party in Café Sperl

The day Walter's Nobel prize was announced I was in Prague receiving a medal from the Czech Academy of Science. At the very moment I came out of the lecture hall where I had been giving a kind of public lecture[3], I received a call on my cellular phone informing me about the staggering news. Of course everybody around me was extremely happy, but they wondered why he was being given the prize in chemistry and not physics.

It turned out that Mara, Walter's wife, would have been unable to attend the ceremony in Stockholm, so Walter decided to postpone his official award ceremony for one year. However, they came to Vienna immediately after Christmas, their first foreign trip after the announcement of the big news. Clearly enough we organized a Nobel prize lecture in one of the main lecture halls of the Technische Universität Wien, but this was not the only kind of official ceremony that took place. There was a formal reception given by the president and another one housed by the president of parliament. In order to make him curious we told Walter several times that by the end of the week there was to be a big surprise party, knowing, of course, that he loves big parties. On that very evening one of his nieces collected him – Mara was already with us – and drove him over to the mysterious place. All the time in the car he impatiently said that he very well knows this neighbourhood. Of course he was right: he was heading for Café Sperl, one of Vienna's most beautiful old coffeehouses, only about 200 meters off the street where he grew up. We had rented the whole café and some 130 people were waiting for Walter. At the party there was not only a very genuine, Viennese-style buffet including dishes like "Schinken- and Krautfleckerln"[4], but also a few surprises waiting for him – for example, a "Maturazeugnis"[5] stating explicitly that he had not passed the Akademisches Gymnasium because of racial reasons. This he accepted with great satisfaction since it was simply the truth. That evening his family, his sister and all his nephews, nieces, grand nephews and grand nieces happily mixed with the rest of the crowd, including quite a few representatives of public life.

---

[3] The title of that lecture was by the way quite funny, namely "Maimonides and Schrödinger's cat".

[4] A type of pasta prepared with either ham ("Schinken") or cabbage ("Kraut").

[5] Certificate for having successfully finished gymnasium.

Peter Weinberger

# The "Walter-Kohn-Zimmer"

When the top floor of the building that houses my group was up for renovation I insisted that a special office be created: the "Walter-Kohn-Zimmer", which would be his office whenever he was in town. This "Zimmer" is not just a plain office: it not only offers a marvellous view of the Habsburg palace, but is also situated just opposite the elementary school that Walter had attended and one can look down on Theobaldgasse, where the Kohn family once lived. This now is the place where he patiently engages himself in long discussions with my students and collaborators, or fights his innumerable emails while being in Vienna. I am pretty sure that occasionally he risks a few glimpses out of the windows, in the direction of his former neighborhood .....

Walter Kohn's first school day in Vienna (photo: Minna Pixner)

# Presenting the "kleine Frau Hofmann" in Zürich

Walter, and in particular Mara, turned out to be especially fond of my non-scientific books. The second one, "Die kleine Frau Hofmann" was an attempt to use contemporary history as background for two "almost criminal stories": in essence it deals with the Nazi era in Vienna and is a kind of "homage" to Walter's and Mara's generation. Since the illustrations in the book are by a quite well-known Swiss sculptor, it was planned right from the beginning to also be presented in Zürich. It just so happened that I knew that Walter would be in Zürich – on one of his annual long visits there – so I asked him to be the presenter. After all, this was a book that had very much to do with the Vienna he knew quite well. Needless to say, Walter did a fantastic job in the Collegium Helveticum, which is housed in the "Semper Sternwarte" and offers a perfect stage for small cultural events. For that book and others, Walter and Mara belong to my group of "test readers": a manuscript that Walter or Mara (or both) likes has a good chance of being accepted in printed form.

## Being His Viennese Secretary

Whenever Walter intends to visit Vienna – on a bigger tour through Europe of course – I first get his schedule, then confusing notes from his secretary and finally a list of people I am supposed to contact and make arrangements such that Walter can meet them while in town. And he loves to stay in the same hotel near his office (the Walter-Kohn-Zimmer) and very near to most of his relatives. As it so happens, that hotel is owned and run by the sister of the chief rabbi of Vienna; he therefore simply refers to it as "The Rabbinic Hotel."[6] He has often assured me that I am the best secretary he has ever had and that he is quite willing to take me over to Santa Barbara as his main secretary. Knowing that he has several other secretaries throughout Europe, I am a little hesitant to accept his offer.

Looking back, from the discovery of the second "K" in KKR in the early 1970s to becoming his local secretary, friend and almost family member, my relationship with Walter has indeed been an experience I wouldn't have wanted to miss. Clearly enough we also had innumerable discussions on physics, from certain properties of Density Functional Theory to transport theory and time-dependent quantum mechanics, but all these discussions were always interwoven with other aspects, such as cultural matters or matters of humanity. Although this now is a quite natural point to stop my recollections of encounters with Walter, I know Walter would have proceded differently, namely by telling a last, a very last story, by pointing out yet another view. And in fact I forgot to mention something very important:

## Walter: Master of Bifurcated Stories

Sholom Aleichem usually starts his stories with one and the same sentence, namely "Damit ich nicht vergesse Ihnen zu erzählen"[7]. Walter just as well

---

[6] The last time Walter scheduled a trip to Vienna, his secretary tried to look up "The Rabbinic Hotel" on the Internet; Of course, she couldn't find an entry.

[7] "So as not to forget to tell you"

could use this phrase since he is the unchallenged master of bifurcated stories, stories that start somewhere, branch off to a different person or situation, eventually returning to the branching point in order to divert again a few moments later. Astonishingly enough he never forgets to complete his stories, including, of course, all the side-stories: his stories always give a complete account of what was to be told, albeit garnished with a few diversions. Questions like "Kennst du (den) ..."[8] are not meant to be answered but are a clear indication for either the beginning of a new master story or just a branching point. If not stopped by external warnings (Mara for example) his stories will unroll to ever-new boundaries: it is simply wonderful to listen to him, the spell of his narrations reaches young and old, independent of their respective cultural backgrounds. Leaving all science aside, for me this is the most loveable side of Walter.

---

[8] "Do you know ..."?

# An Encounter with Walter Kohn in Orsay

Jerry L. Whitten

North Carolina State University, Raleigh, U.S.A.

I met Walter Kohn for the first time in Orsay, France. It was 1974, I think. He was visiting a Solid State CNRS laboratory and I was on sabbatical at CECAM, an organization directed by Carl Moser. I recall a conversation I had with Walter in an office at CECAM. I had been working recently on two things: new configuration methods and formal bounds on electrostatics interactions that occurred in electronic structure problems. We talked about both. I touted the importance of configuration interaction and Walter listened patiently. Obviously, my message didn't get through because his work on density functional theory has gone a long way toward putting CI out of business for many ground state problems. I am not sure I was very convincing on the error bound message either. The idea was to find approximate ways to evaluate the interaction of two densities, $\langle \rho_a(1) \, | \, 1/r_{12} \, | \, \rho_b(2) \rangle$, by expansion methods that required less work than using the exact densities. The positive definite integral $\langle \rho(1) - \rho'(1) \, | \, 1/r_{12} \, | \, \rho(2) - \rho'(2) \rangle \geq 0$ provided the basis for determining such expansions and provided upper and lower bounds on the error of the approximation. When I told Walter about the bound and that the same argument could be used in many other neat ways, practically the first thing he said was, "the bound can't be very good." I tried to argue that the smallness of the error was determined by the closeness of $\rho$ and $\rho'$ and it was the existence of the bound that really mattered. Perhaps I have the last laugh on this one since the expansion method is currently used in several DFT programs.

After moving from Stony Brook to North Carolina State University, I saw Walter again when he visited Duke University and gave a seminar on density functional theory. We had a nice chat at the seminar and he reminded me of our discussion at CECAM almost 20 years earlier.

I am still struggling to keep many-electron CI methods afloat. It's tough to try to match his insight that has led to modern density functional methods. In addition to being the wine connoisseur that Carl Moser says he is, no small accomplishment in itself, the whole world is a connoisseur of his good theory.

Jerry L. Whitten

**About the Author**: Jerry Whitten is Professor of Chemistry at North Carolina State University. His research concerns the chemical physics of surfaces, catalytic reactions on transition metals and photochemistry. His main interests are the development and application of *ab intio* many-electron methods for the accurate description of molecular electronic states and surface processes. Contact: Department of Chemistry, NC State University, Box 8204, Raleigh, NC 27695-8201, U.S.A.; j_whitten@ncsu.edu

# Walter Kohn, World Citizen and Professor Extraordinaire

Henry T. Yang

University of California, Santa Barbara, U.S.A.

How can I find the words to express all that Walter Kohn means to me, and to our University of California, Santa Barbara campus?

Kohn Hall, home of the Kavli Institute for Theoretical Physics, UCSB
(© Kevin Barron)

Walter has been a member of our faculty since 1979, when he became the founding director of the National Science Foundation's Institute for Theoretical Physics at UCSB. The institute's goal was to bring together leading scientists from all over the globe to share their ideas and work together on major topics in theoretical physics and related fields. Under Professor Kohn's dynamic and visionary leadership, it quickly developed into a premier research center that has been looked to as a pioneering model by institutions and countries all over the world. Today, more than 1,000 prominent physicists and scientists from diverse backgrounds participate annually in the institute's programs and conferences.

The institute's unique mode of operation stimulates collaborative interactions that lead to new insights and discoveries at the frontiers of science. Physics Professor Brian Green of Columbia University recently stated, "Theoretical physics would not be where it is today without this vital institute" (now known as the Kavli Institute for Theoretical Physics). World-renowned physicist Stephen Hawking has said, "Some of my best work has been done there." The KITP is a reflection of Professor Kohn's scientific acumen and creative leadership, and our campus is proud that the building that houses the institute is named in his honor.

Nobel Laureate Walter Kohn (© D. Folks)

In 1998, Professor Kohn was awarded the Nobel Prize in Chemistry for his leading role in the development of the density functional theory, now widely used in quantum chemistry. This theory simplifies the calculations to obtain knowledge of the inner structure of molecules, the building blocks of matter. Professor Kohn's discovery meant that scientists could take complex molecules such as proteins and use computer-generated models to predict their chemical behavior. This theory has enabled the development of new drugs, led to a new understanding of the make-up of interstellar matter, and provided insight into how chemical reactions affect our ozone layer, to name just a few examples. I've been told that about half of all publications in quantum chemistry now make reference to his theory.

I remember the headlines of our campus newspaper after Walter Kohn received his Nobel Prize: "Walter Kohn: A Nobelist with Heart." Walter is such a special person, and so beloved on our campus, it made us all tremendously happy to see him recognized with this prize. I said at the time that Walter's inspiration and impact on our proud UCSB campus was beyond what I could describe. That is still true today.

Professor Kohn has received a number of other prestigious awards for his scientific research, including the Oliver Buckley Prize in Solid State Physics, the Davisson–Germer Prize in Surface Physics, the National Medal of Science, the Feenberg Medal in Many-Body Physics, and the Niels Bohr/UNESCO Gold Medal. Clearly, he is one of the world's top scientists. At the same time, he is deeply engaged in matters spiritual and societal.

In 2001, Professor Kohn was the inaugural speaker for the Templeton Research Lectures on Science, Religion, and the Human Experience, sponsored by the Templeton Foundation and hosted by UC Santa Barbara. He spoke about the interaction between science and religion, and about how science and technology pose both great promises and great threats to mankind in this global age.

Rollerblading at West Beach, Santa Barbara (© Vic Cox)

I was particularly struck by his statement that "science by itself is not an adequate basis for conducting one's life." This has certainly been a guiding principle in his own life. Throughout his career, Professor Kohn has been a mentor and role model for colleagues and students alike. Many have been inspired by his incredible life story and his work to promote tolerance and world peace.

Dr. Kohn is a senior statesman, scientist, and advocate for peace whose life and work have had a profound impact on our world. At the same time, he is a down-to-earth person who is great fun to be around. He is most well known to all of us as a patient, kind, and wise man who gives his time generously to help others. His wide-ranging interest in the arts and humanities matches his enthusiasm for the social causes he holds dear. He is an aficionado of classical music and a great scholar whose passion for literature spans centuries and cultures. I also hear he makes a delicious ratatouille. And watch out when Walter gets on his rollerblades!

Walter is deeply devoted to the university and our community. For example, he recently gave the keynote address at the County of Santa Barbara's installation of elected officials and convening of the 2003 Board of Supervisors. To prepare for it, he researched deeply and broadly on the topic, and then he educated us all on the importance of the interdependent and mutually beneficial partnership between UCSB and the community. As another example, Walter has been active in Santa Barbara's Building Bridges Community Coalition, a group dedicated to promoting tolerance and celebrating diversity through collaborative community projects.

Walter Kohn and Henry Yang (© Djamel E. Ramoul)

My wife, Dilling, and I have gotten to know Walter and Mara very well since we came to UCSB in 1994. Their friendship is a gift. The hours we've spent talking and laughing together are ones I treasure dearly. Walter always has such incredible stories to share, full of wisdom, humor, and heart. I value tremendously the guidance and sage advice he has given me over the years.

Two years ago, we presented Walter with the Santa Barbara medal, our campus's highest honor. The inscription read:

"Walter Kohn
Maestro of the molecular,
activist intellectual,
Nobel Laureate, colleague –
UCSB proudly salutes you."

Our campus once again proudly salutes our colleague, friend, teacher, and mentor – Professor Walter Kohn. On behalf of the UCSB community, it is my great honor and delight to congratulate Walter on the very special occasion of his 80th birthday.

**About the Author**: Henry T. Yang is the chancellor of the University of California, Santa Barbara. He is also a professor of mechanical engineering, specializing in aerospace engineering, structural dynamics, composite materials, finite elements, transonic aeroelasticity, wind and earthquake engineering, and manufacturing. He is a member of the National Academy of Engineering and a fellow of the American Institute of Aeronautics and Astronautics. In addition to his administrative duties, Dr. Yang continues to receive federal research grants, guide graduate students, and teach undergraduate courses at UCSB. Contact: Chancellor, University of California, Santa Barbara, CA 93106, U.S.A.; henry.yang@chancellor.ucsb.edu; www.ucsb.edu

# Meeting with Walter Kohn

Weitao Yang

Duke University, Durham, U.S.A.

The first time I met Walter was at Menton in 1991. Bob and Jane Parr most kindly invited Walter and me to dinner before the opening session of the VIIth International Congress of Quantum Chemistry. It was a beautiful summer evening. Bob and Jane were in the restaurant when I got there. Bob then told me that Walter would be elected to the International Academy of Quantum Molecular Science during the opening session in that evening. This was probably when and how Walter became a quantum chemist.

I do not remember much of the conversation during the dinner. What I remember was my mental process from being in awe of Walter's presence at the beginning to feeling at ease in communicating with him by the end of the dinner.

In the 1991 Menton conference, after I gave my talk on the divide-and-conquer method for linear scaling calculations, Kohn told me that my work reminded him the short range of interaction in many-electron systems he had in his mind. I felt that was such a great compliment. Later I had many chances to meet with Walter, often leaving me something to ponder.

During the 6th International Conference on the Applications of Density Functional Theory in Chemistry and Physics in Paris in 1995, knowing that he had spent much time in Paris in the past, I asked Walter what he likes to do and which he likes among the art museums in Paris. He said he enjoys strolling in the Luxembourg Garden and likes the Lady and the Unicorn tapestries in Musée de Cluny. And he recommended that I see the Lady and the Unicorn. Both the Luxembourg Garden and the museum were in the Latin Quarter where the meeting was held. I did both. The garden was just beautiful. The Lady and the Unicorn tapestries were woven in about 1500, six in total representing the six senses – Hearing, Sight, Touch, Smell, Taste, and Love. They are among the most beautiful art treasures of the world. They left me with a lasting impression.

In the summer of 1997, the Symposium on Density Functional Theory and Applications, a Satellite Symposium of the 9th International Congress of Quantum Chemistry, was organized by Mel Levy and myself and held at

Duke University. It was only a week after I just returned from my sabbatical leave at the Hong Kong University of Science and Technology. I really missed the Chinese food in Hong Kong and told Helen, my wife so. Walter was the symposium dinner speaker and Helen and I were sitting by his sides in the head table. After Helen mentioned that I complained about foods in North Carolina, Walter turned to me and said: "Weitao, I think what we are having is very nice food." Later in the after-dinner speech, Walter told the story of the beginning of the density functional theory with P. Hohenberg and Lu Sham.

When Walter received the Nobel Prize in 1998, I was very excited and happy for him. I knew he would get it and said so when I made the case to my colleagues at Duke University to invite Walter to give the London Lecture a few years earlier. I sent Walter a card to congratulate him. A year later, I received a card from Walter with such a kind and encouraging note.

Sept 99

Dear Weitao,
    Finally I find a little time
to say thank you for kind words
last October.
    Think of how much time you
still have!
        Best regards,        Walter

**About the Author**: Weitao Yang is currently a Professor of Chemistry at Duke University. He received his B.S. degree from Peking University and his Ph. D. degree under Professor Robert G. Parr at the University of North Carolina at Chapel Hill. He was a postdoctoral fellow with Professor Parr in Chapel Hill and then with Professor William Miller at the University of California-Berkeley. He joined the chemistry faculty at Duke University in 1989. Dr. Yang's current research interests are in density functional theory, its efficient implementations and its applications to chemistry, biochemistry and nanophysics. Contact: Duke University, Department of Chemistry, Durham, NC 27708, U.S.A.; weitao.yang@duke.edu

# My Experience of Working with Walter

See-Chen Ying

Brown University, Providence, U.S.A.

In the spring of 1969, I was still at Brown University applying for postdoctoral positions when I got a call from Walter Kohn to meet him at the New York March meeting to talk about my application. So I eagerly went down to New York and met Walter for the first time. In those days, Kondo effect was the hottest topic around in condensed matter theory and it seemed like almost any self respecting theorist would want to work on it. So when Walter asked me what I plan to do as a postdoc, I naturally blurted out that one of the topics I want to look at is Kondo effect. It was then that I met the slight head shake and the disapproving wry smile that I would see many times again later. Walter told me that in his opinion, there were too many theorists working on this fashionable problem already while there were still many interesting problems lying around unexplored. So even before I went to work with Walter, I was already taught a valuable lesson. Fortunately, Walter understood the folly of a young budding physicist and decided to offer me the job. Even though I ended up getting a few faculty job offers that year, I eventually decided to accept Walter's offer because I sensed that there was a lot I can learn from him and I was not quite ready to strike out on my own. I have never regretted that decision since.

As a graduate student at Brown working on my Ph.D. thesis under John Quinn, I was given a definite topic and used to working under close guidance and constant discussion with my advisor. While Walter also worked closely with his students, he took a rather different approach to the postdocs, giving us maximum freedom to work on interesting topics in the general area of condensed matter physics. I now understood that this was how Walter trained us to gain independence. At the time, it was a rather nerve-wrecking experience trying to find a good project to work on. I remembered that after dabbling in a small problem in excitonic insulators without making too much advance, Walter suggested that I should take a look at the data that Greg Dash obtained at Seattle which showed power law temperature dependence for the specific heat of a physisorbed layer rather than the exponential form expected. So I started working on this problem, eventually settling on an explanation of the data in terms of commensurate-incommensurate transition within the Frenkel-Kontorowa model. Walter gave me much initial guidance and steered me on the right track. Unfortunately, shortly after the start of the project, Walter took a sabbatical leave to Israel and I was left pretty much on my own. (Those were the days before e-mails!). After he came back to La Jolla, I presented the final manuscript to him describing the results. He read it carefully, gave me an encouraging approval and generously crossed out his name from the authors' list. To this date, I still counted this work as one of my most satisfying papers. Ultimately, being at La Jolla with Walter Kohn, one cannot avoid the pulling force of Density Functional Theory. When John Smith arrived during my second year, we started talking about applying DFT to study chemisorption and eventually coauthored several papers on this with Walter.

One of the most memorable physics experience for me at La Jolla was the Friday Brown Bag Luncheon discussion group that Water organized. Every week, there were informal talks given by a member among the graduate students, postdocs or the visitors. It was characterized by frequent interruptions of questions, exchanges between the speaker and the audience and exchanges among the audiences. There was simply no better way to learn and enjoy physics. The success of this luncheon meeting was largely due to Walter since he can ask probing questions and started a meaningful discussion on just about any topic at the meeting. Whenever Walter was absent, the discussion group usually degenerated into a regular seminar-like talk and became rather dull. Of course, there is a price to pay for this wonderful experience. The speakers are seldom arranged in advance. The postdocs and graduate students, particularly in Walter's group, were all waiting nervously Friday morning in their offices. About an hour or so before the luncheon meeting, Walter would appear and knocked at one of our office doors, and the lucky victim would get the call to talk about what they are working on at the luncheon meeting later. While at the time it was a terrifying experience, later on I realized how much I benefited from organizing my thoughts for the luncheon discussion and getting useful feedbacks.

My final story about Walter was about the time when he came to Brown as a member of a distinguished review committee for the physics Department at Brown University in 1992. As usual, Walter took his responsibility seriously, and besides listening to the presentation of the faculties, also spent some time with a group of graduate students listening to their concerns and opinion about the whole graduate experience. There were no local faculty members at this meeting. Afterwards I asked my graduate student how the meeting went. He said: "Oh, everything went well, only that one student got up and complained about the intermediate solid state physics course (that I was teaching at the time). He said he had to learn something called Density Functional Theory that is not in the textbook and did not seem very useful." Obviously that poor student didn't realize that the father of DFT was on the committee listening to him! I had a good laugh with my student and I am sure that Walter got a big chuckle out of this too.

My experience at La Jolla working with Walter certainly shaped my whole career as a physicist and I am glad I made the right choice. I wish Walter the very best on his $80^{th}$ birthday and hope that he continues to enjoy good health and exert positive influence on the younger generations.

**About the Author**: See-Chen Ying is a Professor of Physics at Brown University, Providence, R.I. His research interests are in condensed matter theory, with special emphasis on the dynamics and equilibrium properties of adsorption systems. Contact: Physics Department, Brown University, Providence, RI 02912, U.S.A.; ying@physics.brown.edu

# Walter Kohn: Mentor and Role Model

Eugene Zaremba

Queen's University, Kingston, Canada

I arrived in La Jolla with my wife Vanda late in December of 1973. I had just completed my PhD with Allan Griffin in Toronto and had secured a NSERC Postdoctoral Fellowship to continue my studies abroad. I was interested in surface physics at the time and I was advised by several people at Toronto that I would do well to take up my postdoc with Walter Kohn at UCSD. I knew little more about density functional theory than what I had heard from Philip Tong in a seminar, but decided to contact Walter, and was delighted to hear that he was willing to take me on as a postdoc. And so our transcontinental journey at the end of 1973 began, a journey to moderate climes and to a new life in physics.

Perhaps it was just poor planning but I prefer to put it down to the naivete and innocence of youth. I arrived at UCSD only to discover that the university was closed for the Christmas holidays. Whether I had Walter's phone number or not I can't recall, but in those days, the thought of contacting a professor at home did not seem proper. And so Vanda and I took up residence at a Travelodge by the ocean to await the start of the new academic term. That was one of our more memorable Christmases, but for all the wrong reasons. To this day we wince at the thought of the Christmas dinner we endured at the Holiday Inn!

Thankfully, the university eventually came to life and I was able to find my place in Urey Hall where I spent the next two and a half years. It's disconcerting that I don't have a clear impression of the first time I met Walter. I suspect I was overwhelmed by the novelty of my surroundings, and perhaps intimidated not only by the faculty but also by the postdocs who all seemed to know much more than I. At least at first, I was on the periphery of Walter's interests and activities. I busied myself writing up my PhD work, attended seminars and tried to learn about what other people were doing. Then at some point, Walter began a graduate course in surface physics and I began to be drawn into his sphere of influence. It was during this course that I first became aware of van der Waals interactions in the context of surfaces. Walter explained how the atom-surface interaction falls off as $Z^{-3}$ and obtained an expression for the van der Waals constant. I was immediately intrigued by the way that this constant could be expressed in terms of an integral over complex frequencies. It was also during this course that I learned about image potentials and the importance of defining a reference plane position. These ideas were later to find application in work we did together.

During this period I also began to appreciate Walter's qualities as a physicist. There is no one I know who seems to enjoy more the very process of explaining things to others, nor is there anyone better at getting ideas across in a simple and thoroughly convincing fashion. Over the years I have come to realize that this is no mean feat. The simplicity and clarity of an exposition belies the depth of thinking that is required to generate it. These impressions have been reinforced by the many talks I have heard Walter give since that time. The main lesson I carried away from these experiences is that physics is tough, but if a subject is understood it can be presented in simple terms. That has been a model I have tried to emulate (but not always successfully). There is no need for buzzwords or jargon which often disguise a superficial understanding. But of course understanding is a very subjective matter. This was brought home to me following a colloquium given by Eugene Wigner at UCSD on the philosophical foundations of quantum mechanics. I was standing next to Walter in his conversation with Wigner when at some point Wigner asked "Do you really understand superconductivity? Because I don't". It was clear that Walter, an expert on the subject, was slightly unnerved by this comment. Obviously Walter did not doubt Wigner's ability to follow the mathematics, but Walter's comfort level with the subject was not shared with this admittedly profound intellect. That was another lesson I learned: there are generational differences and what may be obvious to one is mysterious to another.

During my first six months at UCSD I had only intermittent and sporadic contact with Walter and envied the relationship that others seemed to have with him. This no doubt was more my fault than his and fortunately in the end the problem was resolved. Meeting him in the hall one day, I summoned up my courage and blurted out that my intention in coming to UCSD was to work with *him* and that I had yet to really do so. Walter seemed surprised by this revelation, but recognizing my discomfort, replied that there was no reason why this couldn't happen. He then asked what I wanted to work on and I said atom-surface interactions. That seemed to please him and he indicated that that was a worthwhile project. And so I began to work on a problem that in the end turned out very well. Walter was the ideal mentor, allowing me to pursue my own interests, encouraging me when difficulties arose and guiding me with his depth of understanding. Remarkably, even though this was but one of many problems he was involved in, he provided a continual infusion of insightful comments and ideas.

Surprisingly, although density functional theory was Walter's first love, we never worked together on any aspect of it. However, being in La Jolla, it was impossible to avoid being infected by the bug. It was there that I first began to work on the problem of nonlinear screening in metals with Herb Shore,

Jim Rose and Len Sander. This work continued later with Malcolm Stott at Queen's where we developed the quasi-atom concept which Jens Nørskov independently put forward as effective medium theory and which subsequently led to the embedded atom method of Daw and Baskes. Malcolm and I also considered the problem of linear response within density functional theory, but only the time-independent version which we felt was rigorously based. However the time-dependent version of Zangwill and Soven has proved to be extremely successful, which points out that one should sometimes ignore the strictures of formal rigour. More recently, I have dabbled with density functional based hydrodynamic theories of collective excitations in mesoscopic systems which finds its inspiration in the early work of Ying, Smith and Kohn. Walter's influence may not have been intentional, but it was nevertheless profound.

As a Canadian I cannot help mentioning Walter's various connections to Canada which he himself has proudly proclaimed on several occasions, including his convocation address at Queen's University when he received an honourary degree. As a young man he was interred during the war in a detention camp in Sherbrooke, Quebec. In a recent visit to the university there, he took great pleasure in pointing out that his connection with Sherbrooke predated that of most of his audience! Following the war he studied at the University of Toronto and received a bachelor's degree in a very demanding mathematics and physics program. We as Canadians would like to think that these formative years shaped his future in physics. But whether or not this is true, he himself has shaped a large part of condensed matter physics in Canada. Some, such as Wally Geldart, Sy Vosko (now deceased) and myself, have been touched directly, but many others in both physics and chemistry continue to be influenced, guided and inspired by his body of work. Most of us can only dream of leaving such a legacy.

**About the Author**: Eugene Zaremba is Professor of Physics at Queen's University, Kingston, Canada. He has worked on various aspects of density functional theory, including nonlinear screening in metals, energetics of atoms in solids, linear response theory and collective electronic excitations in mesoscopic systems. His current interest is Bose-Einstein condensation in trapped atomic gases, in particular, the dynamical properties of these systems at finite temperatures. Contact: Department of Physics, Queen's University, Kingston, Ontario, Canada K7L 3N6; zaremba@sparky.phy.queensu.ca

# The Scientist and Human Values

Joseph Zycinski

Archbishop of Lublin, Poland

## Intellectual Solidarity

First time in my life I met Prof. Walter Kohn in Rome, during the celebration of the Holy Jubilee 2000 for university professors. Earlier, when teaching philosophy of physics at the Pontifical Academy of Cracow, I met him only through his scientific publications concerning condensed matter theory. In his biography, I read when Prof. Kohn received Nobel Prize in 1998, I found that his mother was born in Brody, Galicia. This news attracted my attention because a part of Galicia belonged to that part of southern Poland where I spent seven years as Roman Catholic Bishop of Tarnow in 1990–1997. At that time I found there many relics of Jewish culture. To save most important of them, together with Sigmunt Nissenbaum and my Jewish friends, I helped to organize the restoring of the synagogue in Bobowa, the centre well-known of the spiritual activity of the famous Jewish tzadik. Consequently, when during the Holy Jubilee celebrations inspired by John Paul II, we met with Prof. Kohn at the scientific panel in Rome, he seemed to me already a friendly person rooted in the culture so close to my pastoral concern.

This impression was strengthened during the days of our meeting at the University of Tor Vergata. The gentle style of Prof. Kohn, his intellectual openness to many discussed issues, his warm friendliness and subtle reactions to the proposed alternative interpretations created an atmosphere in which human values were present not only in our debates but also in our practice. On the last day of celebrations, on September 10, 2000, an encounter with John Paul II was planned. I met Prof. Kohn in the group of scientists which was to be introduced personally to the Pope. As a paradigmatic scientist, he seemed a little bit embarrassed by all protocol procedures, being afraid that bureaucratic regulations could be more important in such meetings than the basic human solidarity. When I noticed his embarrassment, I told him that I know well the Pope and that there is no reason to introduce any formal elements in this expression of intellectual solidarity. The Pope – I commented – for many years worked as the university professor at the Catholic University of Lublin. Thus there is no reason to worry how one university professor meets another one.

As a matter of fact there was no reason at all. When I watched them at their special encounter, both John Paul II and Prof. Kohn seemed to express the same basic solidarity in human values which has been so important for scientists as well as for the people of the Church. Their radiating faces confirmed the opinion that the new form of cooperation between the Church and the University is needed to express a common care of basic human values. To foster this kind of intellectual solidarity, the Pope systematically invites to his summer residence in Castel Gandolfo many scientists open to interdisciplinary dialogue. When inviting them to this enriching dialogue, John Paul II wrote: "Science can purify religion from error and superstition; religion can purify science from idolatry and false absolutes. Each can draw the other into a wider world, a world in which both can flourish. ... Both the Church and the scientific community ... shall make our choices much better if we live in a collaborative interaction."[1]

## Overcoming the Separation

The split between modern science and theology was based on methodological foundations. Galileo, when developing his new science, was right in eliminating any reference to angels from physical explanation of planetary motions. Though he never denied the value of theological explanation, the author of the *Dialogo* was right when he claimed that all references to theological factors must be excluded from the domain of scientific research. He was right because if, in the spirit of medieval astronomy, one were to refer to the role of angels to explain the motion of planets, one could always introduce the hypothesis of angels to explain any set of empirical data. As a result, in such an approach, astronomy would merely remain a branch of applied angelology where reference to angels would be enough to explain any physical phenomenon.

Galileo with his methodology of the new science contributed to the rise of the so called principle of methodological positivism, underlying the growth of modern science. This principle does not permit the scientist to refer to any extra-natural factors to explain physical processes. For methodological reasons, when the chain of causal dependencies is examined on the level of the natural sciences, one can introduce in this chain neither radical discontinuity by assuming a special divine intervention nor a sequence of micro-discontinuities in which God's presence would be described in the language of information increase. Certainly, the principle at stake is methodological in nature. It does not preclude the existence of an extra-physical agent which

---

[1] John Paul II, Letter to George Coyne SI on 300 anniversary of publishing Newton's *Principia*, in: *Physics, Philosophy, Theology. A Common Quest for Understanding*, ed. R. J. Russell et al., Vatican Observatory 1988, M 13.

can be referred to in philosophical research. It only brackets his role in the cognitive framework of natural sciences.

Our belief in epistemological Galilean economy of explanation resulted in the well-known principle of Ockham's razor. This very principle, however, is methodological in nature, not doctrinal. It could inspire effective research procedure, but it cannot provide simple answers to complicated philosophical questions. Even on the level of physical research, this principle often played a heuristically negative role. Its critics indicate many examples of the disadvantageous consequences of its application in science. It is true that in the 19th century the appeal to Ockham's razor retarded the development of extragalactic astronomy by nearly a century. Dogmatic adherents to the Ockham's principle argued at that time that there are no extragalactic objects because all observed astronomical phenomena can be explained more economically by reference to the objects in our Galaxy. This search for simplicity resulted in a false cosmological model. As a result, in the contemporary philosophy of science, a special "de-Ockhamization" program has been promoted in which the Ockham's principle has a relative, not an absolute value.

Galileo's methodology resulted in eliminating from science any reference to God's presence in nature or to the importance of human values. In the picture of nature provided by pre-Galilean texts we find quite different elements. The biblical description of natural processes contained in the *Psalms*, presents God clothed in majesty and splendor, wearing the light as a robe (Ps 103, 1–4). In this perspective there is a harmony between the world of nature and that of spirituality. This harmony has been constituted by great cedars of Lebanon and also by tiny herbs, mountains full of marmots and wild goats, and also by the Spirit of God which renews the face of the earth (Ps 103). Specific aspects of this harmony are seen in the Gospel, in which the unseen divine reality reveals itself in some of the basic elements of nature – the lilies of the field and the vine plant, the fig tree and the storm on the lake, the Bethlehem plain and the Garden of Olives. In the biblical perspective, as in the philosophical reflections of Plato and Leibniz, God reveals his presence not in the gaps of our knowledge about nature but in the harmony of nature. A particular form of this tradition would be developed in the 20th century by Alfred N. Whitehead in his process philosophy. The co-author of the *Pricipia Mathematica*, compares the role of God, immanent in his creation, to the creativity of the Poet of the world.

Poetry has not been appreciated in scientific theories by physicists themselves. The question of separation between poetry and science returns in a new form in contemporary methodological debates. Many problems that seemed poetic and meaningless for the generation the Vienna Circle, in the 1930's, appear important and meaningful in the context of the present debates on the unified physical theory. One cannot avoid here the question as to why the language of mathematics has been so effective in the physical description of nature and why physical processes are described by universal physical laws even when they could have been nothing but an uncoordinated mess. These questions could be regarded as the counterpart of the classical philosophical problem: "Why does being exist when there could have existed mere nothingness?" This question, criticised as trivial, poetic and meaningless by empirical positivists in the 1930's, now can be expressed in a new form which is meaningful also for empiricists: "Why at all are there mathematically described laws of physics when nature could have existed as an uncoordinated chaos?"

Consequently, we should affirm that it is impossible to be a scientist, even an atheist scientist, and not be struck by the awesome beauty of mathematical formulae in physical theories, by the mystery of harmony and of ingenuity in nature. The most impressive expression of this harmony can be found in the existence of the underlying mathematical order. This very order, described by abstract mathematical formulae, has often been regarded as a bare fact or an unintelligible mystery. In my opinion, to explain the nature of such an order one has to go beyond mathematics as well as beyond the natural sciences. This transcending of the level of scientific discovery brings us to the divine Logos underlying the mathematical structure of the world.

There are authors who call such a structure "the rationality field", "the formal field" or the "matrix of the universe". Jan Łukasiewicz, the well-known representative of the Polish School of logic, argued that the reality of ideal mathematical structures independent of human experience could be regarded as an expression of God's presence in nature. Regardless our terminological preferences, in the aesthetic component of nature we find the factor which inspires our amazement and directs our attention toward theological and philosophical aspects of nature. They are consistent with Galileo's version of methodological positivism. The theologians and the scientists must not be in conflict but, following different methodological rules, they could look for an integral vision of nature.

## Searching for a New Solidarity

A colleague of mine, Prof. Michael Heller, who is both an expert in quantum cosmology and a Roman Catholic priest, during a public conference was introduced as a member of the Vatican Astronomical Observatory. On the next day, press informed that the scholarly discussion was dominated by a member of the staff of an *Astrological* Observatory. For the young generation of journalists, who never had a chance to study astronomy, astrology seems closer to everyday life while Harry Potter replaces Albert Einstein in demonstrating physical effects which seem contrary to our common sense expectations. This expression of postmodern mentality discloses the necessity of a closer cooperation between the Academia and the Church. The appreciation of rational arguments, the recognition of objective truths and meaningful theories should unite us to overcome the temptation of nihilism or cynicism, so characteristic of contemporary postmodern culture.

Joseph Zycinski

As a Catholic bishop I accepted all arguments presented during the Rome meeting by Prof. Kohn, a physicist brought up in Jewish milieu. Going beyond his physical arguments, I claim that, in accordance with physical theories, God's presence can be discovered in the various forms of harmony: physical, aesthetic, mathematical, ethical, spiritual. The different expressions of this harmony constitute the domain of human ecology, which facilitates our personal growth thanks to the continuous cooperation with the divine Poet of the world. In this dynamic framework, all of us are invited to multiply the beauty of existence while spiritual harmony becomes an important component of our human ecology.

New cultural challenges brought by important cultural transformations require our common answer. It is our duty to bring new optimism to the people who suffer of frustration and despair. What remains a main challenge in the actual process of globalization, it is its false anthropology. In the last century, various versions of false anthropologies resulted in two totalitarian systems. Now, in a pragmatic-commercial approach, one tries to transform human community into consumers' society. A crisis of contemporary humanism seems to be a consequence of such a praxis. This crisis began with the proud Nietzsche's declaration of the death of God; one century later, in much more humble statements, the death of man or of the end of human subject were announced. To bring the intellectual counterproposal to such statements, we must be harbingers of hope. In our global culture we need universal values to share the dramas of particular human person. These universal values can be offered neither by politics nor by business; we should look for them in scientific milieus as well as in theological and philosophical debates.

There was a time when universal values were denied on ideological basis. The defenders of the so called model of *homo sovieticus* claimed that both principles of ethics and principles of logic are so specific in the Soviet Union that it is impossible to refer to Western standards in the appraisal of human life in the East. After the fall of totalitarian systems, we could recognize the importance of basic human values when we celebrate the jubilees of scientists who dedicate their life to truth and who try to build basic solidarity rooted in human values.

**About the Author**: Joseph Zycinski, born in 1948, professor of philosophy at the Catholic University of Lublin, Poland; Archbishop of Lublin; Grand Chancellor of the Catholic University of Lublin. Member of the European Academy of Science and Art; Member of the Pontifical Council for Culture; Member of the Russian Academy of Natural Sciences; author of nearly 40 books in philosophy of science, relativistic cosmology, history of the relationship between natural sciences and Christian faith. Contact: ul. Prymasa Stefana Wyszynskiego 2, PL-20950 Lublin (Poland); j.zycinski@kuria.lublin.pl; http://www.kuria.lublin.pl/biskup/

# Appendix I – Autobiography

Walter Kohn

University of California, Santa Barbara, U.S.A.

I suppose I am not the first Nobelist who, on the occasion of receiving this Prize, wonders how on earth, by what strange alchemy of family background, teachers, friends, talents and especially accidents of history and of personal life he or she arrived at this point. I have browsed in previous volumes of "Les Prix Nobel" and I know that there are others whose eventual destinies were foreshadowed early in their lives – mathematical precocity, champion bird watching, insatiable reading, mechanical genius. Not in my case, at least not before my late teens. On the contrary: An early photo of my older sister and myself, taken at a children's costume party in Vienna – I look about 7 years old – shows me dressed up in a dark suit and a black top hat, toy glasses pushed down my nose, and carrying a large sign under my arm with the inscription "Professor Know-Nothing".

Here then is my attempt to convey to the reader how, at age 75, I see my life which brought me to the present point: a long-retired professor of theoretical physics at the University of California, still loving and doing physics, including chemical physics, mostly together with young people less than half my age; moderately involved in the life of my community of Santa Barbara and in broader political and social issues; with unremarkable hobbies such as listening to classical music, reading (including French literature), walking with my wife Mara or alone, a little cooking (unjustifiably proud of my ratatouille); and a weekly half hour of relaxed roller blading along the shore, a throwback to the ice-skating of my Viennese childhood. My three daughters and three grandchildren all live in California and so we get to see each other reasonably often.

I was naturalized as an American citizen in 1957 and this has been my primary self-identity ever since. But, like many other scientists, I also have a strong sense of global citizenship, including especially Canada, Denmark, England, France and Israel, where I have worked and lived with a family for considerable periods, and where I have some of my closest friends.

My feelings towards Austria, my native land, are – and will remain – very painful. They are dominated by my vivid recollections of 1 1/2 years as a

Walter as "Professor Know-Nothing" with his older sister, about 1930/31

Jewish boy under the Austrian Nazi regime, and by the subsequent murder of my parents, Salomon and Gittel Kohn, of other relatives and several teachers, during the holocaust. At the same time I have in recent years been glad to work with Austrians, one or two generations younger than I: Physicists, some teachers at my former High School and young people (Gedenkdiener) who face the dark years of Austria's past honestly and constructively.

On another level, I want to mention that I have a strong Jewish identity and – over the years – have been involved in several Jewish projects, such as the establishment of a strong program of Judaic Studies at the University of California in San Diego.

My father, who had lost a brother, fighting on the Austrian side in World War I, was a committed pacifist. However, while the Nazi barbarians and their collaborators threatened the entire world, I could not accept his philosophy and, after several earlier attempts, was finally accepted into the Canadian Infantry Corps during the last year of World War II. Many decades later I became active in attempts to bring an end to the US-Soviet nuclear arms race and became a leader of unsuccessful faculty initiatives to terminate the role of the University of California as manager of the nuclear weapons laboratories at Los Alamos and Livermore. I offered early support to Jeffrey Leiffer, the founder of the student Pugwash movement which concerns itself with global issues having a strong scientific component and in which scientists can play a useful role. Twenty years after its founding this organization continues strong and vibrant. My commitment to a humane and peaceful world continues to this day. I have just joined the Board of the Population Institute because I am convinced that early stabilization of the world's population is important for the attainment of this objective.

Walter Kohn with his sister and parents, about 1932

After these introductory general reflections from my present vantage point I would now like to give an idea of my childhood and adolescence. I was born in 1923 into a middle class Jewish family in Vienna, a few years after the end of World War I, which was disastrous from the Austrian point of view. Both my parents were born in parts of the former Austro-Hungarian Empire, my father in Hodonin, Moravia, my mother in Brody, then in Galicia, Poland, now in the Ukraine. Later they both moved to the capital of Vienna along with their parents. I have no recollection of my father's parents, who died relatively young. My maternal grandparents Rappaport were orthodox Jews who lived a simple life of retirement and, in the case of my grandfather, of prayer and the study of religious texts in a small nearby synagogue, a Schul as it was called. My father carried on a business, Postkartenverlag Brueder Kohn Wien I, whose main product was high quality art postcards, mostly based on paintings by contemporary artists which were commissioned by his firm. The business had flourished in the first two decades of the century but then, in part due to the death of his brother Adolf in World War I, to the dismantlement of the Austrian monarchy and to a worldwide economic depression, it gradually fell on hard times in the 1920s and 1930s. My father struggled from crisis to crisis to keep the business going and to support the family. Left over from the prosperous times was a wonderful summer property in Heringsdorf at the Baltic Sea, not far from Berlin, where my mother, sister and I spent our summer vacations until Hitler came to power in Germany in 1933. My father came for occasional visits (The firm had a branch in Berlin). My mother was a highly educated woman with a good knowledge of German, Latin, Polish and French and some acquaintance with Greek, Hebrew and English. I believe that she had completed an academically oriented High School in Galicia. Through her parents we maintained contact with traditional Judaism. At the same time my parents, especially my father, also were a part of the secular artistic and intellectual life of Vienna.

Walter Kohn and his mother in 1937

After I had completed a public elementary school, my mother enrolled me in the Akademische Gymnasium, a fine public high school in Vienna's inner city. There, for almost five years, I received an excellent education, strongly oriented toward Latin and Greek, until March 1938, when Hitler Germany annexed Austria. (This so-called Anschluss was, after a few weeks, supported by the great majority of the Austrian population). Until that time my favorite subject had been Latin, whose architecture and succinctness I loved. By contrast, I had no interest in, nor apparent talent for, mathematics which was routinely taught and gave me the only C in high school. During this time it was my tacit understanding that I would eventually be asked to take over the family business, a prospect which I faced with resignation and without the least enthusiasm.

The Anschluss changed everything: The family business was confiscated but my father was required to continue its management without any compensation; my sister managed to emigrate rather promptly to England; and I was expelled from my school.

In the following fall I was able to enter a Jewish school, the Chajes Gymnasium, where I had two extraordinary teachers: In physics, Dr. Emil Nohel, and in mathematics Dr. Victor Sabbath. While outside the school walls arbitrary acts of persecution and brutality took place, on the inside these two inspired teachers conveyed to us their own deep understanding and love of their subjects. I take this occasion to record my profound gratitude for their inspiration to which I owe my initial interest in science. (Alas, they both became victims of Nazi barbarism).

I note with deep gratitude that twice, during the Second World War, after having been separated from my parents who were unable to leave Austria, I was taken into the homes of two wonderful families who had never seen me before: Charles and Eva Hauff in Sussex, England, who also welcomed my older sister, Minna. Charles, like my father, was in art publishing and they had a business relationship. A few years later, Dr. Bruno Mendel and his wife Hertha of Toronto, Canada, took me and my friend Joseph Eisinger into their family. (They also supported three other young Nazi refugees). Both of these families strongly encouraged me in my studies, the Hauffs at the East Grinstead County School in Sussex and the Mendels at the University of Toronto. I cannot imagine how I might have become a scientist without their help.

My first wife, Lois Kohn, gave me invaluable support during the early phases of my scientific career; my present wife of over 20 years, Mara, has supported me in the latter phases of my scientific life. She also created a wonderful home for us, and gave me an entire new family, including her father Vishniac, a biologist as well as a noted photographer of pre-war Jewish communities in Eastern Europe, and her mother Luta. (They both died rather recently, well into their nineties).

After these rather personal reminiscences I now turn to a brief description of my life as a scientist.

When I arrived in England in August 1939, three weeks before the outbreak of World War II, I had my mind set on becoming a farmer (I had seen too many unemployed intellectuals during the 1930s), and I started out on a training arm in Kent. However, I became seriously ill and physically weak with meningitis, and so in January 1940 my "acting parents", the Hauffs, arranged for me to attend the above-mentioned county school, where – after a period of uncertainty – I concentrated on mathematics, physics and chemistry.

However, in May 1940, shortly after I had turned 17, and while the German army swept through Western Europe and Britain girded for a possible German air-assault, Churchill ordered most male "enemy aliens" (i.e., holders of enemy passports, like myself) to be interned ("Collar the lot" was his crisp order). I spent about two months in various British camps, including the Isle of Man, where my school sent me the books I needed to study. There I also audited, with little comprehension, some lectures on mathematics and physics, offered by mature interned scientists.

In July 1940, I was shipped on, as part of a British convoy moving through U-boat-infested waters, to Quebec City in Canada; and from there, by train, to a camp in Trois Rivieres, which housed both German civilian internees and refugees like myself. Again various internee-taught courses were offered. The one which interested me most was a course on set-theory given by the mathematician Dr. Fritz Rothberger and attended by two students. Dr. Rothberger, from Vienna, a most kind and unassuming man, had been an advanced private scholar in Cambridge, England, when the internment order was issued. His love for the intrinsic depth and beauty of mathematics was gradually absorbed by his students.

Later I was moved around among various other camps in Quebec and New Brunswick. Another fellow internee, Dr. A. Heckscher, an art historian, organized a fine camp school for young people like myself, whose education had been interrupted and who prepared to take official Canadian High School exams. In this way I passed the McGill University junior Matriculation exam and exams in mathematics, physics and chemistry on the senior matriculation level. At this point, at age 18, I was pretty firmly looking forward to a career in physics, with a strong secondary interest in mathematics.

I mention with gratitude that camp educational programs received support from the Canadian Red Cross and Jewish Canadian philanthropic sources. I also mention that in most camps we had the opportunity to work as lumberjacks and earn 20 cents per day. With this princely sum, carefully saved up, I was able to buy Hardy's Pure Mathematics and Slater's Chemical Physics, books which are still on my shelves. In January 1942, having been cleared by Scotland Yard of being a potential spy, I was released from internment and welcomed by the family of Professor Bruno Mendel in Toronto. At this point I planned to take up engineering rather than physics, in order to be able to support my parents after the war. The Mendels introduced me to Professor Leopold Infeld who had come to Toronto after several years with Einstein. Infeld, after talking with me (in a kind of drawing room oral exam), concluded that my real love was physics and advised me to major in an excellent, very stiff program, then called mathematics and physics, at the University of Toronto. He argued that this program would enable me to earn a decent living at least as well as an engineering program.

However, because of my now German nationality, I was not allowed into the chemistry building, where war work was in progress, and hence I could not enroll in any chemistry courses. (In fact, the last time I attended a chemistry class was in my English school at the age of 17). Since chemistry was required, this seemed to sink any hope of enrolling. Here I express my deep appreciation to Dean and head of mathematics, Samuel Beatty, who helped me, and several others, nevertheless to enter mathematics and physics as special students, whose status was regularized one or two years later.

I was fortunate to find an extraordinary mathematics and applied mathematics program in Toronto. Luminous members whom I recall with special vividness were the algebraist Richard Brauer, the non-Euclidean geometer, H.S.M. Coxeter, the aforementioned Leopold Infeld, and the classical applied mathematicians John Lighton Synge and Alexander Weinstein. This group had been largely assembled by Dean Beatty. In those years the University of Toronto team of mathematics students, competing with teams from the leading North-American Institutions, consistently won the annual Putman competition. (For the record I remark that I never participated). Physics too had many distinguished faculty members, largely recruited by John C. McLennan, one of the earliest low temperature physicists, who had died before I arrived. They included the Raman specialist H.L. Welsh, M.F. Crawford in optics and the low-temperature physicists H.G. Smith and A.D. Misener. Among my fellow students was Arthur Schawlow, who later was to share the Nobel Prize for the development of the laser.

Walter Kohn as a Canadian soldier, 1944/45

During one or two summers, as well as part-time during the school year, I worked for a small Canadian company which developed electrical instruments for military planes. A little later I spent two summers, working for a geophysicist, looking for (and finding!) gold deposits in northern Ontario and Quebec.

After my junior year I joined the Canadian Army. An excellent upper division course in mechanics by A. Weinstein had introduced me to the dynamics of tops and gyroscopes. While in the army I used my spare time to develop new strict bounds on the precession of heavy, symmetrical tops. This paper, "Contour Integration in the Theory of the Spherical Pendulum and the Heavy Symmetrical Top" was published in the Transactions of American Mathematical Society. At the end of one year's army service, having completed only 2 1/2 out of the 4-year undergraduate program, I received a war-time bachelor's degree "on-active-service" in applied mathematics.

In the year 1945–46, after my discharge from the army, I took an excellent crash master's program, including some of the senior courses which I had missed, graduate courses, a master's thesis consisting of my paper on tops and a paper on scaling of atomic wave-functions.

My teachers wisely insisted that I do not stay on in Toronto for a Ph.D, but financial support for further study was very hard to come by. Eventually I was thrilled to receive a fine Lehman fellowship at Harvard. Leopold Infeld recommended that I should try to be accepted by Julian Schwinger, whom he knew and who, still in his 20s, was already one of the most exciting theoretical physicists in the world.

Arriving from the relatively isolated University of Toronto and finding myself at the illustrious Harvard, where many faculty and graduate students had just come back from doing brilliant war-related work at Los Alamos, the MIT Radiation Laboratory, etc., I felt very insecure and set as my goal survival for at least one year. The Department Chair, J.H. Van Vleck, was very kind and referred to me as the Toronto-Kohn to distinguish me from another person who, I gathered, had caused some trouble. Once Van Vleck told me of an idea in the band-theory of solids, later known as the quantum defect method, and asked me if I would like to work on it. I asked for time to consider. When I returned a few days later, without in the least grasping his idea, I thanked him for the opportunity but explained that, while I did not yet know in what subfield of physics I wanted to do my thesis, I was sure it would not be in solid state physics. This problem then became the thesis of Thomas Kuhn, (later a renowned philosopher of science), and was further developed by myself and others. In spite of my original disconnect with Van Vleck, solid state physics soon became the center of my professional life and Van Vleck and I became lifelong friends.

After my encounter with Van Vleck I presented myself to Julian Schwinger requesting to be accepted as one of his thesis students. His evident brilliance as a researcher and as a lecturer in advanced graduate courses (such as waveguides and nuclear physics) attracted large numbers of students, including many who had returned to their studies after spending "time out" on various war-related projects.

I told Schwinger briefly of my very modest efforts using variational principles. He himself had developed brilliant new Green's function variational principles during the war for wave-guides, optics and nuclear physics (Soon afterwards Green's functions played an important role in his Nobel-Prize-winning work on quantum electrodynamics). He accepted me within minutes as one of his approximately 10 thesis students. He suggested that I should try to develop a Green's function variational method for three-body scattering problems, like low-energy neutron-deuteron scattering, while warning me ominously, that he himself had tried and failed. Some six months later, when I had obtained some partial, very unsatisfactory results, I looked for alternative approaches and soon found a rather elementary formulation, later known as Kohn's variational principle for scattering, and useful for nuclear, atomic and molecular problems. Since I had circumvented Schwinger's beloved Green's functions, I felt that he was very disappointed. Nevertheless he accepted this work as my thesis in 1948. (Much later L. Fadeev offered his celebrated solution of the three-body scattering problem).

My Harvard friends, close and not so close, included P.W. Anderson, N. Bloembergen, H. Broida (a little later), K. Case, F. De Hoffman, J. Eisenstein, R. Glauber, T. Kuhn, R. Landauer, B. Mottelson, G. Pake, F. Rohrlich, and C. Slichter. Schwinger's brilliant lectures on nuclear physics also attracted many students and Postdocs from MIT, including J. Blatt, M. Goldberger, and J.M. Luttinger. Quite a number of this remarkable group would become lifelong friends, and one – J.M. "Quin" Luttinger – also my closest collaborators for 13 years, 1954–66. Almost all went on to outstanding careers of one sort or another.

I was totally surprised and thrilled when in the spring of 1948 Schwinger offered to keep me at Harvard for up to three years. I had the choice of being a regular post-doctoral fellow or dividing my time equally between research and teaching. Wisely – as it turned out – I chose the latter. For the next two years I shared an office with Sidney Borowitz, later Chancellor of New York University, who had a similar appointment. We were to assist Schwinger in his work on quantum electrodynamics and the emerging field theory of strong interactions between nucleons and mesons. In view of Schwinger's deep physical insights and celebrated mathematical power, I soon felt almost completely useless. Borowitz and I did make some very minor contributions, while the greats, especially Schwinger and Feynman, seemed to be on their way to unplumbed, perhaps ultimate depths.

For the summer of 1949, 1 got a job in the Polaroid laboratory in Cambridge, Mass., just before the Polaroid camera made its public appearance. My task was to bring some understanding to the mechanism by which charged particles falling on a photographic plate lead to a photographic image. (This technique had just been introduced to study cosmic rays). I therefore needed to learn something about solid state physics and occasionally, when I encountered things I didn't understand, I consulted Van Vleck.

It seems that these meetings gave him the erroneous impression that I knew something about the subject. For one day he explained to me that he was about to take a leave of absence and, "since you are familiar with solid state physics", he asked me if I could teach a course on this subject, which he had planned to offer. This time, frustrated with my work on quantum field theory, I agreed. I had a family, jobs were scarce, and I thought that broadening my competence into a new, more practical, area might give me more opportunities.

So, relying largely on the excellent, relatively recent monograph by F. Seitz, "Modern Theory Of Solids", I taught one of the first broad courses on Solid State Physics in the United States. My "students" included several of my friends, N. Bloembergen, C. Slichter and G. Pake who conducted experiments (later considered as classics) in the brand-new area of nuclear magnetic resonance which had just been opened up by E. Purcell at Harvard and F. Bloch at Stanford. Some of my students often understood much more than I, they were charitable towards their teacher.

At about the same time I did some calculations suggested by Bloembergen, on the recently discovered, so-called Knight shift of nuclear magnetic resonance, and, in this connection, returning to my old love of variational methods, developed a new variational approach to the study of wavefunctions in periodic crystals.

Although my appointment was good for another year and a half, I began actively looking for a more long-term position. I was a naturalized Canadian citizen, with the warmest feelings towards Canada, and explored every Canadian university known to me. No opportunities presented themselves. Neither did the very meager US market for young theorists yield an academic offer. At this point a promising possibility appeared for a position in a new Westinghouse nuclear reactor laboratory outside of Pittsburgh. But during a visit it turned out that US citizenship was required and so this possibility too vanished. At that moment I was unbelievably lucky. While in Pittsburgh, I stayed with my Canadian friend Alfred Schild, who taught in the mathematics department at the Carnegie Institute of Technology (now Carnegie Mellon University). He remarked that F. Seitz and several of his colleagus had just left the physics department and moved to Illinois, so that – he thought – there might be an opening for me there. It turned out that the Department Chair, Ed Creutz was looking rather desperately for somebody who could teach a course in solid state physics and also keep an eye on the graduate students who had lost their "doctor-fathers". Within 48 hours I had a telegram offering me a job!

A few weeks later a happy complication arose. I had earlier applied for a National Research Council fellowship for 1950–51 and now it came through. A request for a short postponement was firmly denied. Fortunately, Ed Creutz agreed to give me a one-year leave of absence, provided I first taught a compressed course in solid state physics. So on December 31, 1950 (to satisfy the terms of my fellowship) I arrived in Copenhagen.

Originally I had planned to revert to nuclear physics there, in particular the the the structure of the deuteron. But in the meantime I had become a solid state physicist. Unfortunately no one in Copenhagen, including Niels Bohr, had even heard the expression "Solid State Physics". For a while I worked on old projects. Then, with an Indian visitor named Vachaspati (no initial), I published a criticism of Froehlich's pre-BCS theory of superconductivity, and also did some work on scattering theory.

In the spring of 1951, I was told that an expected visitor for the coming year had dropped out and that the Bohr Institute could provide me with an Oersted fellowship to remain there until the fall of 1952. Very exciting work was going on in Copenhagen, which eventually led to the great "Collective Model of the Nucleus" of A. Bohr and B. Mottelson, both of whom had become close friends. Furthermore my family and I had fallen in love with Denmark and the Danish people. A letter from Niels Bohr to my department chair at Carnegie quickly resulted in the extension of my leave of absence till the fall of 1952.

In the summer of 1951, I became a substitute teacher, replacing an ill lecturer at the first summer school at Les Houches, near Chamonix in France, conceived and organized by a dynamic young French woman, Cécile Morette De Witt. As an "expert" in solid state physics, I offered a few lectures on that subject. Wolfgang Pauli, who visited, when he learned of my meager knowledge of solids, mostly metallic sodium, asked me, true to form, if I was a professor of physics or of sodium. He was equally acerbic about himself. Some 50 years old at the time, he described himself as "a child-wonder in menopause" ("ein Wunderkind in den Wechseljahren"). But my most important encounter was with Res Jost, an assistant of Pauli at the ETH in Zurich, with whom I shared an interest in the so-called inverse scattering problem: given asymptotic information, (such as phase-shifts as function of energy), of a particle scattered by a potential $V(r)$, what quantitative information can be inferred about this potential? Later that year, we both found ourselves in Copenhagen and addressed this problem in earnest. Jost, at the time a senior fellow at the Institute for Advanced Study in Princeton, had to return there before we had finished our work. A few months later, in the spring of 1952, I received an invitation from Robert Oppenheimer, to come to Princeton for a few weeks to finish our project. In an intensive and most enjoyable collaboration, we succeeded in obtaining a complete solution for S-wave scattering by a spherical potential. At about the same time I.M. Gel'fand in the Soviet Union published his celebrated work on the inverse problem. Jost and I remained close lifelong friends until his death in 1989.

After my return to Carnegie Tech in 1952, I began a major collaboration with N. Rostoker, then an assistant of an experimentalist, later a distinguished plasma theorist. We developed a theory for the energy band structure of electrons for periodic potentials, harking back to my earlier experience with scattering, Green's functions and variational methods. We showed how to determine the bandstructure from a knowledge of purely geometric structure constants and a small number ($\sim 3$) of scattering phase-shifts of the potential in a single sphericalized cell. By a different approach this theory was also obtained by J. Korringa. It continues to be used under the acronym KKR. Other work during my Carnegie years, 1950–59, includes the image of the metallic Fermi Surface in the phonon spectrum (Kohn anomaly); exponential localization of Wannier functions; and the nature of the insulating state.

My most distinguished colleague and good friend at Carnegie was G.C. Wick, and my first PhD's were D. Schechter and V. Ambegaokar. I also greatly benefitted from my interaction with T. Holstein at Westinghouse.

In 1953, with support from Van Vleck, I obtained a summerjob at Bell Labs as assistant of W. Shockley, the co-inventor of the transistor. My project was radiation damage of Si and Ge by energetic electrons, critical for the use of the recently developed semiconductor devices for applications in outer space. In particular, I established a reasonably accurate energy threshold for permanent displacement of a nucleus from its regular lattice position, substantially smaller than had been previously presumed. Bell Labs at that time was without question the world's outstanding center for research in solid state physics and for the first time, gave me a perspective over this fascinating, rich field. Bardeen, Brattain and Shockley , after their invention of the transistor, were the great heroes. Other world class theorists were C. Herring, G. Wannier and my brilliant friend from Harvard, P.W. Anderson. With a few interruptions I was to return to Bell Labs every year until 1966. I owe this institution my growing up from amateur to professional.

In the summer of 1954 both Quin Luttinger and I were at Bell Labs and began our 13-year long collaborations, along with other work outside our professional "marriage". (Our close friendship lasted till his death in 1997). The all-important impurity states in the transistor materials Si and Ge, which govern their electrical and many of their optical properties, were under intense experimental study, which we complemented by theoretical work using so-called effective mass theory. In 1957, I wrote a comprehensive review on this subject. We (mostly Luttinger) also developed an effective Hamiltonian in the presence of magnetic fields, for the complex holes in these elements. A little later we obtained the first non-heuristic derivation of the Boltzman transport equation for *quantum mechanical* particles. There followed several years of studies of many-body theories, including Luttinger's famous one-dimensional "Luttinger liquid" and the "Luttinger's theorem" about the conservation of the volume enclosed by a metallic Fermi surface, in the presence of electron electron interaction. Finally, in 1966, we showed that superconductivity occurs even with purely repulsive interactions – contrary to conventional wisdom and possibly relevant to the much later discovery of high-$T_c$ superconductors.

In 1960, when I moved to the University of California San Diego, California, my scientific interactions with Luttinger, then at Columbia University, and with Bell Labs gradually diminished. I did some consulting at the nearby

General Atomic Laboratory, interacting primarily with J. Appel. My university colleagues included G. Feher, B. Maple, B. Matthias, S. Schultz, H. Suhl and J. Wheatley, – a wonderful environment. During my 19-year stay there I typically worked with two postdocs and four graduate students. A high water mark period were the late 1960s, early 1970s, including N. Lang, D. Mermin, M. Rice, L.J. Sham, D. Sherrington, and J. Smith.

I now come to the development of density functional theory (DFT). In the fall of 1963, I spent a sabbatical semester at the École Normale Supérieure in Paris, as guest and in the spacious office of my friend Philippe Nozières. Since my Carnegie days I had been interested in the electronic structure of alloys, a subject of intense experimental interest in both the physics and metallurgy departments. In Paris I read some of the metallurgical literature, in which the concept of the effective charge $e^*$ of an atom in an alloy was prominent, which characterized in a rough way the transfer of charge between atomic cells. It was a local point of view in coordinate space, in contrast to the emphasis on *delocalized* waves in *momentum space*, such as Bloch-waves in an average periodic crystal, used for the rough description of substitutional alloys. At this point the question occurred to me whether, in general, an alloy is *completely* or only partially characterized by its electronic density distribution $n(r)$: In the back of my mind I knew that this was the case in the Thomas-Fermi approximation of interacting electron systems; also, from the "rigid band model" of substitutional alloys of neighboring elements, I knew that there was a 1-to-1 correspondence between a weak perturbing potential $\delta v(r)$ and the corresponding small change $\delta n(r)$ of the density distribution. Finally it occurred to me that for a single particle there is an explicit elementary relation between the potential $v(r)$ and the density, $n(r)$, of the groundstate. Taken together, these provided strong support for the conjecture that the density $n(r)$ completely determines the external potential $v(r)$. This would imply that $n(r)$ which integrates to $N$, the total number of electrons, also determines the total Hamilton $H$ and hence all properties derivable from $H$ and $N$, e.g. the wavefunction of the 17th excited state, $\Psi^{17}(r_1, \ldots r_N)$! Could this be true? And how could it be decided? Could two different potentials, $v_1(r)$ and $v_2(r)$, with associated different groundstates $\Psi_1(r_1, \ldots r_N)$ and $\Psi_2(r_1, \ldots r_N)$ give rise to the *same* density distribution? It turned out that a simple 3-line argument, using my beloved Rayleigh–Ritz variational principle, confirmed the conjecture. It seemed such a remarkable result that I did not trust myself.

By this time I had become friends with another inhabitant of Nozière's office, Pierre Hohenberg, a lively young American, recently arrived in Paris after a one-year fellowship in the Soviet Union. Having completed some work there he seemed to be "between" problems and I asked if he would be interested in joining me. He was. The first task was a literature search to see if this simple result was already known; apparently not. In short order we had recast the Rayleigh–Ritz variational theorem for the groundstate energy in terms of the density $n(r)$ instead of the many electron wave function $\Psi$, leading to what is now called the Hohenberg–Kohn (HK) variational principle. We fleshed out this work with various approximations and published it.

Shortly afterwards I returned to San Diego where my new postdoctoral fellow, Lu J. Sham had already arrived. Together we derived from the HK variational principle what are now known as the Kohn-Sham (KS) equations, which have found extensive use by physicists and chemists, including members of my group.

Since the 1970s I have also been working on the theory of surfaces, mostly electronic structure. The work with Lang in the early 1970s, using DFT, picked up and carried forward where J. Bardeen's thesis had left off in the 1930s.

In 1979, I moved to the University of California, Santa Barbara to become the initial director of the National Science Foundation's Institute for Theoretical Physics (1979–84). I have continued to work with postdoctoral fellows and students on DFT and other problems that I had put aside in previous years. Since the middle 1980s, I have also had increasing, fruitful interactions with theoretical chemists. I mention especially Robert Parr, the first major theoretical chemist to believe in the potential promise of DFT for chemistry who, together with his young co-workers, has made major contributions, both conceptual and computational.

Since beginning this autobiographical sketch I have turned 76. I enormously enjoy the continuing progress by my younger DFT colleagues and my own collaboration with some of them. Looking back I feel very fortunate to have had a small part in the great drama of scientific progress, and most thankful to all those, including family, kindly "acting parents", teachers, colleagues, students, and collaborators of all ages, who made it possible.

From Les Prix Nobel 1998 (© The Nobel Foundation 1998) plus photos added by us

# Appendix II – A Musical CV

## GRADUS AD PARNASSUM

*The Musical Curriculum Vitae of Walter Kohn*
*on the occasion of his Sixtieth Birthday*

When Jim Langer asked if I could say a few after-dinner words at the banquet at Walter Kohn's 60th birthday conference, held March 1983 in Santa Barbara, I said no, but if he could supply me with a copy of Walter's CV, I'd do my best to set it to music. Instead of a polite "OK, I'll get somebody else" he sent a CV by return mail, leaving me to make good on my bluff. Walter's Musical CV follows the tune and much of the sense – insofar as there is any – of Sir Joseph Porter's well-known song, "When I was a lad", from Act I of Gilbert and Sullivan's operetta H.M.S. Pinafore, with its famous refrain "now he is the Ruler of the Queen's navee!" Vinay Ambegaokar and I sang it, while I provided the piano accompaniment and Vinay energetically conducted the audience, who lustily sang the part of the chorus, belting out parenthetical commentaries from distributed texts. To my regret our own copies vanished long ago. I attempted, not entirely accurately, to reconstruct a few fragments while writing my contribution to this volume, leading me to include in my essay a plea to any reader who happened to have retained a libretto from the party to send it on to me. Within days of the appearance of my essay on the secret pre-publication website, Sudip Chakravarty wrote to say that he had kept his text for 20 years and would be glad to send me a copy. So after two decades I became reacquainted with my song. Many of its phrases are lifted from Walter Kohn papers, but I'm proud to say that I am the sole author of the rhyme at the beginning of Verse VII.

*David Mermin*

# I

When I was a lad I thought a lot
About the heavy and symmetric top.
I thought so hard they published me
In the 'Merican Math'matical Society.
(The American Math'matical Society.)
But Mathematics weren't for me
So I went to get a Harvard Physics Ph.D.
(No Mathematics wasn't for he
So instead he went to get a Physics Ph.D.)

# II

In Cambridge Town I made my way
To where the Schwinger students laugh and play
And found it recreational
To investigate some methods variational.
(He began to ponder methods variational.)
My trial functions soon taught me
All the facts about collisions of light nucle-ee.
(Those trial functions soon taught he
Everything about colliding little nucle-ee.)

# III

Those Schwinger students I had seen
Preoccupied themselves with functions Green,
But functions Green were grey to me
So I said Goodbye to quantum field theory.
(Yes, he said Farewell to quantum field theory.)
Oh no more nucleons for me:
I prefer to think about cohesive energy.
(For nucleons do not bring glee
When a lad can think about cohesive energy!)

# IV

So Pittsburgh next became my home
And soon my intellect began to roam
To èlectronic properties
Like donor levels near impurities.
(And acceptor levels near impurities.)
Amongst the phonons I could see
The image of the surface of the Fermi sea!
(Amongst the phonons who but he
Would think to seek the image of the Fermi sea?)

### V

But when La Jolla called to me
I left the Institute of Carnegie.
And there I came to understand
That one need not calculate each energy band.
(No, one doesn't have to figure out each energy band.)
For èlectronic properties
Are universal functions of their densities.
(Yes every single property
Is a functional of electronic density!)

### VI

Inhòmogeneous èlectrons
Are at surfaces and lie along the valence bonds.
Whatever question you may ask
I've a functional designed to do that very task.
(Yes his functional works brilliantly at every task.)
It works so well they appointed me
To be the first Director of the ITP.
(From èlectronic density
He went on to be Director of the ITP.)

### VII

Today I live in Santa Barbara
As the reigning physics theory Czar.
The Journey there was hard and long:
My Curriculum Vitae makes a seven verse song!
(His Curriculum Vitae needs a very long song.)
From a lad with a top I have grown to be
The one and only Leader of the ITP.
(From a heavy top with symmetry
He has risen to Director of the ITP.)

Printing: Mercedes-Druck, Berlin
Binding: Stein+Lehmann, Berlin